Dynamic Modeling and Econometrics in Economics and Finance

Volume 14

Editors
Stefan Mittnik
University of Munich
Munich, Germany

Willi Semmler
Bielefeld University
Bielefeld, Germany
and
New School for Social Research
New York, USA

For further volumes:
www.springer.com/series/5859

Jesús Crespo Cuaresma • Tapio Palokangas •
Alexander Tarasyev
Editors

Green Growth and Sustainable Development

Springer

Editors
Jesús Crespo Cuaresma
Department of Economics
Vienna University of Economics
and Business
Vienna, Austria

Tapio Palokangas
University of Helsinki and HECER
Helsinki, Finland

Alexander Tarasyev
Dynamic Systems
Institute of Mathematics and Mechanics,
UrB RAS
Ekaterinburg, Russia
and
Advanced System Analysis
International Institute for Applied Systems
Analysis (IIASA)
Laxenburg, Austria

ISSN 1566-0419
ISBN 978-3-642-44874-4 ISBN 978-3-642-34354-4 (eBook)
DOI 10.1007/978-3-642-34354-4
Springer Heidelberg New York Dordrecht London

Softcover reprint of the hardcover 1st edition 2013

Printed on acid-free paper

Springer is part of Springer Science+Business Media (www.springer.com)

Introduction

This book examines problems associated with green growth and sustainable development using economic theory, systems theory and optimal control theory. Especially, questions of sustainability of economic growth are crucial for green economies envisaging local and global environmental constraints, biodiversity management, human capital development, improvements in resource productivity, and investments in new technologies. The focus is on elaboration of series of models which catch interactions of production factors as driving forces of economic growth.

The book is based on material provided for the Symposium "Green Growth and Sustainable Development Symposium" held at the International Institute for Applied Systems Analysis (IIASA) on the 9th–10th of December, 2011, within the IIASA Project "Driving Forces of Economic Growth" (ECG). The symposium was organized by coordinators of the ECG project: Jesus Crespo-Cuaresma from IIASA World Population Program, Tapio Palokangas and Alexander Tarasyev from IIASA Advanced Systems Analysis Program.

The chapters of the book are presented in a popular-science style and can be interesting to a wide range of scientific literature readers. The prime audience for the book is economists, environmental managers, mathematicians and engineers working on problems of economic growth and environment regulation. The mathematical part of the book is written in a rigorous manner, and the detailed analysis is expected to be of interest to specialists in dynamic systems, optimal control and applications to economic modeling.

The book starts with three chapters on optimal economic growth with environmental constraints, continues with three chapters on the control of climate change, abatement and biodiversity, and with two chapters on dynamics of environmental policy, and ends up with two applications of systems theory to the supply of energy. As a whole, the book provides an integrated view on environmental policy in the setting of economic growth and dynamics.

The first part, "Optimal Economic Growth with an Environmental Constraint", is devoted to methodological problems of economic growth arising in presence of environmental constraints and lack of resources. It considers models oriented on optimization of investments in production factors, examines optimal transitional dynamics and forecasts future trends of economic development.

Sergey Aseev, Konstantin Besov and Serguei Kaniovski study optimal research and extraction policies in an endogenous growth model in which both production and research use exhaustible resources. They show that optimal growth is not sustainable if the accumulation of knowledge uses exhaustible resources, or if the returns to scale in research are decreasing, or the economy is too small. The authors state the model as an infinite-horizon optimal control problem with an integral constraint on the control variables. They consider the main mathematical aspects of the problem, establish an existence theorem and derive an appropriate version of the Pontryagin maximum principle, and give a complete characterization of the optimal transitional dynamics.

Ulla Lehmijoki develops a long-run consumer optimization model where pollution aggravates mortality. In that model, the optimal growth path is sustainable if it provides non-decreasing consumption for a non-decreasing population. As usually, optimality and sustainability may conflict; with population endogenous to pollution, this conflict may ultimately lead the human species toward self-imposed extinction. In that case, not even technical progress can warrant sustainability.

Alexander Tarasyev and Bing Zhu analyse a dynamic optimization model of investment in the improvement of resource productivity in order to find balanced growth trends in terms of consumption and the use of natural resources. This research is closely connected with the problems concerning shortages of natural resource stocks, the security of supply of energy and materials, and the environmental effectiveness of their consumption. The author's main idea is to introduce an integrated environment to control the management of the investment process in the development of basic production factors such as capital, energy and material consumption. Essential features of the model are (i) the possibility to invest in economy's dematerialization and (ii) the price formation mechanism which presumes the rapid growth of prices on exhausting materials. The authors solve optimal control problem for the investment process by the Pontryagin maximum principle, showing that for specific range of the model parameters, there exists the unique steady state of the Hamiltonian system. This enables the existence of a sustainable growth path in an economy with exhausting resources. By these results, strategies for investment in dematerialization, resource and environmental management can be constructed for the purpose to shift the economic system from non-optimal paths to sustainable development.

The second part, "Biodiversity, Abatement and Climate Change", comprises chapters that address application of dynamic systems modeling to economic growth, with especial focus on issues of environmental impact and policy regulation.

Tapio Palokangas examines a group of countries where the conservation of land anywhere yields utility everywhere through biodiversity. All countries produce the same good from labor and land and improve their productivity through abatement investment. The international agency performing biodiversity management is self-interested. Palokangas compares three cases of biodiversity management: (i) laissez-faire, (ii) the regulation of land use, and (iii) subsidies to the conservation of land.

His results are the following. Regulation promotes biodiversity, abatement and welfare. Because subsidies must be financed by distortionary taxes, the replacement of regulation by subsidies hampers biodiversity, abatement and welfare. Applied to NATURA 2000 in the EU, this suggests that regulation without any budget is the appropriate degree of authority for the Commission.

Due to recent global discussions about climate change and its possible consequences, the usage of environmental policy instruments with the intent to counteract against the current environmental developments has become increasingly important. Elke Moser, Alexia Prskawetz and Gernot Tragler investigate the impact of environmental standards on capital accumulation and R&D investments in an economy where both, brown (dirty) as well as green (clean) capital can be used in production. Environmental regulation as policy instrument is commonly supposed to reduce or ideally minimize emissions and pollution. The authors show that such regulations can repress innovation and economic growth rather than induce a shift toward a greener technology.

Nordhaus (2000, 2008) developed a dynamic model that links economic growth with climate change. Helmut Maurer, Johann Jacob Preuß and Willi Semmler present variants of that model, building on the dynamic model of Greiner et al. (2010), who discuss multiple equilibria and thresholds in a canonical optimal control problem with infinite horizon. The authors study various extensions of the basic optimal control problem and compare the solutions for finite horizon and infinite horizon. They admit terminal constraints for the state variable, consider the impacts of constraints (such as CO2 and temperature constraints) on abatement policies and consumption, and attempt to control the temperature by suitable penalties on the temperature. The introduction of these constraints allows exploring the implications for mitigation policies that arise from the Kyoto treaty (CO2 constraint) and the Copenhagen agreement (temperature constraint).

The third part, "Dynamics of Environmental Policy with an Oligopoly", deals with dynamic game modeling of oligopolistic competition with environmental externalities.

Luca Lambertini and George Leitmann adopt a stepwise approach to the analysis of a dynamic oligopoly game in which production exploits a natural resource and pollutes the environment. They start with simple models where firms' output is not a function of the natural resource to end up with a full-fledged model in which (i) the resource is explicitly considered as an input of production and (ii) the natural resource and pollution interact via the respective state equations. They show that the relationship between the welfare properties of the economic system and the intensity of competition is sensitive to the degree of accuracy with which the model is constructed.

The established view on oligopolistic competition with environmental externalities has it that, since firms neglect the external effect, their incentive to invest in R&D for pollution abatement is nil unless they are subject to some form of environmental taxation. Davide Dragone, Luca Lambertini and Arsen Palestini take a dynamic approach to this issue, showing by a simple differential game that conclusion reached by the static literature is not robust: the introduction of dynamics shows that firms do invest in R&D for environmental-friendly technologies throughout the game, as long as R&D is accompanied by an output restriction exhibiting a distinctively collusive flavor. The authors show that there exists a feasible tax rate that induces profit-seeking firms to choose a combination of output and R&D that imposes the same level of social welfare as in the first best.

The fourth part, "Application of Dynamic Systems to Energy Supply", consists of papers that analyse of the role of energy supply and new technologies in economic growth and energy transition.

Chihiro Watanabe and Jae-Ho Shin consider green technology driven energy for sustainable growth by the Japanese example. Japan has constructed a sophisticated co-evolutionary dynamism between innovation and institutional systems by transforming external crises into a springboard for new innovation. This can largely be attributed to the unique features of the nation such as having a strong motivation to overcoming fear based on xenophobia and uncertainty avoidance as well as abundant curiosity, assimilation proficiency, and thoroughness in learning and absorption. Such explicit dynamism was typically demonstrated by technology substitution for energy in the 1970s leading Japan to achieve a high-technology miracle in the 1980s. While this dynamism shifted to the opposite direction in the 1990s due to a system conflict with the rise of the information society, recent increase in oil prices has signaled the possibility of a paradigm shift to a post-oil society. In addition, global economic stagnation due to excessive consumption has been inducing "new normal" customers supra-functionality. The authors show by empirical analysis that these trends inevitably compel to explore high efficient photovoltaic (PV) system.

Bo Hu, Armin Leopold and Stefan Pickl present a System Dynamics model that depicts the development of the energy market in Germany in an aggregated form. They use that model to compare different possible pathways of the impeding energy transition. Their simulations show that a concept presented by the German Advisory Council on Environment (SRU) will only achieve about 31% GHG mitigation in 2025 compared to 1990, despite the high costs due to planned huge storage capacity. A more effective GHG mitigation of about 40% can be achieved at a lower cost by making use of higher wind and photovoltaic capacities in combination with the capability to produce synthetic natural gas (SNG) using excess electricity from wind and solar energy.

We expect that the results of this monograph provide readers with a methodological technique and modeling environment for analysis of economic growth processes, and give an instrument for forecasting growth trends and improving its precision. Furthermore, the models elaborated in the book could serve as helpful tools for policy advice in designing strategies of economic and environmental management.

Contents

Contributors

Sergey Aseev International Institute for Applied Systems Analysis (IIASA), Laxenburg, Austria; Steklov Mathematical Institute, Moscow, Russia

Konstantin Besov Steklov Mathematical Institute, Moscow, Russia

Davide Dragone Department of Economics, University of Bologna, Bologna, Italy

Bo Hu Universität der Bundeswehr Munich, Neubiberg, Germany

Serguei Kaniovski Austrian Institute of Economic Research (WIFO), Vienna, Austria

Luca Lambertini Department of Economics, University of Bologna, Bologna, Italy; ENCORE, University of Amsterdam, Amsterdam, The Netherlands

Ulla Lehmijoki University of Helsinki and HECER, Helsinki, Finland

George Leitmann College of Engineering, University of California at Berkeley, Berkeley, CA, USA

Armin Leopold Universität der Bundeswehr Munich, Neubiberg, Germany

Helmut Maurer Institut für Numerische und Angewandte Mathematik, Universität Muenster, Münster, Germany

Elke Moser Wittgenstein Centre (IIASA, VID/ÖAW, WU), VID/ÖAW, Vienna, Austria

Arsen Palestini MEMOTEF, Sapienza University of Rome, Rome, Italy

Tapio Palokangas University of Helsinki and HECER, Helsinki, Finland

Stefan Pickl Universität der Bundeswehr Munich, Neubiberg, Germany

Johann Jakob Preuß Institut für Numerische und Angewandte Mathematik, Universität Muenster, Münster, Germany

Alexia Prskawetz Wittgenstein Centre (IIASA, VID/ÖAW, WU), VID/ÖAW, Vienna, Austria; Research Group Economics, Institute of Mathematical Methods in Economics, Vienna University of Technology, Vienna, Austria

Willi Semmler Economics Department, The New School for Social Research, New York, NY, USA

Jae-Ho Shin Korea Institute of S&T Evaluation and Planning (KISTEP), Seoul, Korea

Alexander Tarasyev IMM UrB RAS, Ekaterinburg, Russia; IIASA, Laxenburg, Austria

Gernot Tragler Research Group ORCOS, Institute of Mathematical Methods in Economics, Vienna University of Technology, Vienna, Austria

Chihiro Watanabe Department of Industrial Management, Tokyo Seitoku University, Tokyo, Japan; Tokyo Institute of Technology, Tokyo, Japan; National University of Singapore, Singapore, Singapore; University of Jyväskylä, Jyväskylä, Finland; International Institute for Applied Systems Analysis (IIASA), Laxenburg, Austria

Bing Zhu IIASA, Laxenburg, Austria; Tsinghua University, Beijing, China

Part I
Optimal Economic Growth with an Environmental Constraint

The Problem of Optimal Endogenous Growth with Exhaustible Resources Revisited

Sergey Aseev, Konstantin Besov, and Serguei Kaniovski

1 Introduction

Endogenous growth theory identifies technological progress as a means of sustaining economic growth despite the reliance on exhaustible resources as inputs to production. The supply of an exhaustible resource may limit growth, unless the economy can either substitute away from the resource or increase the efficiency of the resource's use to offset its scarcity. Can an optimal research and extraction policy compensate the negative effects on production (consumption) that arise due to scarcity of the exhaustible resource?

Existing literature in the tradition of Dasgupta-Heal-Solow-Stiglitz (Dasgupta and Heal 1974; Stiglitz 1974) offers an affirmative answer in a scenario in which production requires the resource but the accumulation of knowledge does not. Our point of departure is the 'toy economy' model by Charles Jones (Jones 2004), as one of the simplest models of endogenous growth. We show that resource-dependency may preclude perpetual growth along a welfare-maximizing output trajectory if technical progress depends on the resource, or, as was advocated by Jones (1999, 2004), technical progress shows weak scale effects, or the economy is too small. The

S. Aseev
International Institute for Applied Systems Analysis (IIASA), Schlossplatz 1, 2361 Laxenburg, Austria

S. Aseev · K. Besov
Steklov Mathematical Institute, Gubkina 8, 119991 Moscow, Russia

S. Aseev
e-mail: aseev@mi.ras.ru

K. Besov
e-mail: kbesov@mi.ras.ru

S. Kaniovski (✉)
Austrian Institute of Economic Research (WIFO), Arsenal, Objekt 20, 1103 Vienna, Austria
e-mail: serguei.kaniovski@wifo.ac.at

J. Crespo Cuaresma et al. (eds.), *Green Growth and Sustainable Development*, Dynamic Modeling and Econometrics in Economics and Finance 14,
DOI 10.1007/978-3-642-34354-4_1,

possibility of the research sector being dependent on an exhaustible resource challenges the feasibility of perpetual growth, while the strong scale condition seems less profound in view of many existing models that obtain balanced positive growth without it, see Segerstrom (1998), Young (1998). The minimum size condition is the least restrictive of all conditions.

We thus show that welfare-maximizing growth can be either perpetual or transient, and derive optimal research and extraction policies in each scenario. Perpetual growth is balanced and the optimal research policy allocates a constant fraction of the labor force to research. Perpetual growth is feasible even in the absence of population growth and is thus fully-endogenous. Perpetual growth becomes unfeasible if technical progress requires the resource or has weak scale effects. In this more realistic scenario it is optimal to pursue a certain ratio of the knowledge to the resource stock. In the resulting 'rise and decline' scenario output grows initially but stagnates and eventually declines following stagnation of the knowledge stock. In either scenario it is optimal to deplete the resource according to the well-known Hotelling rule.

The model is formulated as an infinite-horizon optimal control problem whose solution is a welfare-maximizing dynamic research and extraction policy. The model includes an integral constraint (in L^1-space) associated with a finite stock of an exhaustible resource. Such integral constraints on control (policy) variables are the defining feature of a class of models in the resource and growth literature (see examples in Weitzman 2003).

Due to the unbounded nature of controls corresponding to the extraction policy, we cannot directly appeal to the standard results on existence of an optimal control in the class of locally bounded measurable functions (such results usually rely on pointwise boundedness conditions; see, for example, Cesari 1983). We overcome this difficulty by reducing our problem to one without integral constraints. This allows us to prove an existence result and apply a version of the Pontryagin maximum principle for problems with a dominating discount developed in Aseev and Kryazhimskiy (2004), Aseev and Kryazhimskii (2007) to fully characterize the optimal transitional dynamics.

The infinite time horizon gives rise to specific mathematical features of the Pontryagin maximum principle (Pontrjagin et al. 1964). The most characteristic feature is that the standard transversality conditions may fail (see examples in Aseev and Kryazhimskii 2007; Halkin 1974; Shell 1969). There exist modifications of the Pontryagin maximum principle that pay attention to this phenomenon (Aseev and Kryazhimskiy 2004; Aseev and Kryazhimskii 2007; Aubin and Clarke 1979; Benveniste and Scheinkman 1982; Seierstad and Sydsæter 1987). Yet our problem fails to satisfy the assumptions imposed in them due to the integral constraint.

When the Hamiltonian is concave, some infinite horizon problems can be solved by means of well-known sufficient conditions (Arrow and Kurz 1970; Seierstad and Sydsæter 1987). This is a standard way of solving many optimal economic growth problems (see Acemoglu 2008). Nevertheless, even in our simple model the concavity of the Hamiltonian cannot be asserted for all relevant parameter values.

In this paper we follow the more general approach based on necessary conditions and an existence theorem. It should be stressed that without an existence theorem one cannot be sure that a path satisfying the necessary conditions exists, or that one of the paths satisfying the necessary conditions is indeed a solution (see the discussion in Romer 1986).

2 The Model

In the following two-sector endogenous growth model the production sector yields output that is consumed, while the research sector augments the productivity of the production means. Both sectors require an exhaustible resource as an input. There are constant returns to scale in production, and either weak or strong scale effects in the research sector.

At every instant $t \in [0, \infty)$, the economy produces output $Y(t) > 0$, which is assumed to be described by a Cobb–Douglas production function:

$$Y(t) = A(t)^{\varkappa}\left[L - L^A(t)\right]^{\alpha} R_1(t)^{1-\alpha} \quad \text{where } \alpha \in (0, 1) \text{ and } \varkappa > 0. \tag{1}$$

Here $A(t) > 0$ is the current knowledge stock and $R_1(t) > 0$ is the quantity of the exhaustible resource used in production. The population (total labor supply) is fixed at $L > 0$. Part of the labor $L - L^A(t)$ is employed in production, while the other part $L^A(t) \in [0, L)$ is allocated to research.

The amount of new knowledge produced at time t depends on the hitherto accumulated knowledge, the number of researchers and the portion of the exhaustible resource used in research:

$$\dot{A}(t) = A(t)^{\theta}\left[L^A(t)\right]^{\eta} R_2(t)^{1-\eta} \quad \text{where } \eta \in (0, 1] \text{ and } \theta \in (0, 1]. \tag{2}$$

Here $R_2(t) \geq 0$ is the quantity of the exhaustible resource used in research; typically $R_2(t)$ is small compared to $R_1(t)$. The initial knowledge stock is given by $A(0) = A_0 > 0$. If $\theta \in (0, 1)$, then growth rate of the knowledge stock decreases while the knowledge stock expands. The case of $\theta < 0$—when the expansion of knowledge is progressively more difficult—has also been considered in the literature (see, e.g., Jones 1999). Empirical evidence supports the idea of weak scale effects, i.e. $\theta < 1$, in the production of knowledge. We retain $\theta = 1$ as a special case of strong scale effects.

The fact that the stock of the exhaustible resource is finite imposes the following integral constraint on the controls $R_1(\cdot)$ and $R_2(\cdot)$:

$$\int_0^{\infty} \left[R_1(t) + R_2(t)\right] dt \leq S_0, \tag{3}$$

where $S_0 > 0$ is the initial supply of the exhaustible resource.

The welfare is measured by a discounted logarithmic utility function, maximizing which amounts to maximizing future growth rates. This leads to the following objective functional for the economy (see (1)):

$$\begin{aligned} J\big(A(\cdot), L^A(\cdot), R_1(\cdot)\big) &= \int_0^\infty e^{-\rho t}\big\{\ln\big[Y(t)\big]\big\}dt \\ &= \int_0^\infty e^{-\rho t}\big\{\varkappa \ln A(t) + \alpha \ln\big[L - L^A(t)\big] + (1-\alpha)\ln R_1(t)\big\}dt, \end{aligned}$$

where $\rho > 0$ is a subjective discount rate.

Given the parameters $\theta \in (0, 1]$, $\alpha \in (0, 1)$, $\varkappa > 0$, $\eta \in (0, 1]$, $\rho > 0$, $L > 0$ and $S_0 > 0$, the optimization problem $J(A(\cdot), L^A(\cdot), R_1(\cdot)) \to \max$, subject to (2) and the resource constraint (3), can be formulated as the following infinite-horizon optimal control problem (P):

$$\dot{A}(t) = A(t)^\theta \big[L^A(t)\big]^\eta R_2(t)^{1-\eta}, \tag{4}$$

$$L^A(t) \in [0, L), \qquad R_1(t) > 0, \qquad R_2(t) \geq 0,$$

$$\int_0^\infty \big[R_1(t) + R_2(t)\big]dt \leq S_0, \tag{5}$$

$$A(0) = A_0 > 0, \tag{6}$$

$$\begin{aligned} J\big(A(\cdot), L^A(\cdot), R_1(\cdot)\big) = \int_0^\infty & e^{-\rho t}\big\{\varkappa \ln A(t) + \alpha \ln\big[L - L^A(t)\big] \\ &+ (1-\alpha)\ln R_1(t)\big\}dt \\ &\to \max. \end{aligned} \tag{7}$$

The above formulation follows closely the model suggested in Sect. 5.2.1 by Groth (2006) who characterized the steady-state solution. In this paper we offer a rigorous derivation of the optimal solution, which will be a steady state only under certain 'knife-edge' conditions that are unlikely to hold. These conditions are: the exhaustible resource is not an input to the production of knowledge ($\eta = 1$), and the accumulation of knowledge has strong scale effects ($\theta = 1$) (see Sect. 6).

The main difference between our model and the model by Groth lies in the exclusion of capital as a third factor input to production and a perpetual inventory equation describing the evolution of the capital stock. A closer look of what essentially is Groth's model in Cabo et al. (2010) shows that knowledge and capital accumulation together lead (under some conditions) to explosive growth. Reaching infinite output in finite time is not a reasonable feature for an infinite time horizon growth model. We therefore exclude capital accumulation in our present model, planning to return to the issue of the interplay of capital and knowledge accumulation in future work.

Next, we introduce the basic elements of model in the terminology of optimal control theory. By an *admissible control* $w(\cdot) : [0, \infty) \to \mathbb{R}^3$ in problem (P) we mean a triple $w(\cdot) = (L^A(\cdot), R_1(\cdot), R_2(\cdot))$, $t \geq 0$, of (locally) bounded measurable functions $L^A(\cdot)$, $R_1(\cdot)$ and $R_2(\cdot)$ each of which is defined on the infinite half-open time interval $[0, \infty)$ and satisfies the respective constraints in (5).

An *admissible trajectory* $A(\cdot) : [0, \tau) \to \mathbb{R}^1$, $\tau > 0$, corresponding to an admissible control $w(\cdot)$ is a (locally) absolutely continuous function $A(\cdot)$ which is a (Carathéodory) solution (see Filippov 1988) of the differential equation (4) on some (finite or infinite) time interval $[0, \tau)$, subject to the initial condition (6).

Due to (4) and the integral constraint in (5), for any admissible control $w(\cdot) = (L^A(\cdot), R_1(\cdot), R_2(\cdot))$ the corresponding admissible trajectory $A(\cdot)$ can be extended to the whole infinite interval $[0, \infty)$. Consequently, in what follows, without loss of generality, we always assume that any admissible trajectory $A(\cdot)$ is defined on $[0, \infty)$.

A pair $(A(\cdot), w(\cdot))$, where $w(\cdot)$ is an admissible control and $A(\cdot)$ is the corresponding admissible trajectory, is called an *admissible pair* (or a *process*) in problem (P).

For any admissible pair $(A(\cdot), w(\cdot))$ the improper integral in (7) converges either to $-\infty$ or to a finite real. Moreover, it is uniformly bounded from above; i.e., there is a number $M \geq 0$ such that

$$\sup_{(A(\cdot),w(\cdot))} \int_0^\infty e^{-\rho t}\big\{\varkappa \ln A(t) + \alpha \ln\big[L - L^A(t)\big] + (1-\alpha)\ln R_1(t)\big\}dt \leq M, \quad (8)$$

where the supremum is taken over all admissible pairs $(A(\cdot), w(\cdot))$.

Indeed, due to the integral constraint in (5), for any admissible control $w(\cdot)$ we have

$$\int_0^\infty e^{-\rho t} \ln R_1(t)dt < \int_0^\infty e^{-\rho t} R_1(t)dt < S_0. \quad (9)$$

Further, for an arbitrary admissible trajectory $A(\cdot)$ we have

$$A(t)^\theta \leq A(t) + 1, \quad t \geq 0.$$

Then, due to (4), we obtain

$$\frac{d}{dt}\ln\big(A(t)+1\big) = \frac{\dot{A}(t)}{A(t)+1} \leq L^\eta R_2(t)^{1-\eta}, \quad t \geq 0,$$

and hence

$$\begin{aligned} \ln\big(A(t)+1\big) &\leq \ln(A_0+1) + L^\eta \int_0^t R_2(s)^{1-\eta}ds \\ &\leq \ln(A_0+1) + L^\eta \int_0^t \big(1+R_2(s)\big)ds \\ &< \ln(A_0+1) + L^\eta(t+S_0), \quad t \geq 0. \end{aligned} \quad (10)$$

This inequality immediately implies the following inequality for an arbitrary admissible trajectory $A(\cdot)$:

$$\int_0^\infty e^{-\rho t}\ln A(t)dt < \int_0^\infty e^{-\rho t}\ln\big(A(t)+1\big)dt < \frac{\ln(A_0+1)+L^\eta S_0}{\rho}+\frac{L^\eta}{\rho^2}. \quad (11)$$

Since $L^A(t)\in[0,L)$, $t\geq 0$ (see (5)), inequalities (9) and (11) provide the following uniform estimate for all control processes $(A(\cdot), w(\cdot))$:

$$\int_0^\infty e^{-\rho t}\big\{\varkappa\ln A(t)+\alpha\ln\big[L-L^A(t)\big]+(1-\alpha)\ln R_1(t)\big\}dt$$
$$<\varkappa\frac{\ln(A_0+1)+L^\eta S_0}{\rho}+\frac{\varkappa L^\eta}{\rho^2}+\frac{\alpha\ln L}{\rho}+(1-\alpha)S_0.$$

This furnishes the proof of inequality (8).

The uniform bound (8) allows us to define an *optimal control* $w_*(\cdot):[0,\infty)\to\mathbb{R}^3$ in problem (P) as a welfare-maximizing triple $w_*(\cdot)=(L_*^A(\cdot), R_{1*}(\cdot), R_{2*}(\cdot))$ of dynamic labor and extraction policies adopted in the research and production sectors. The corresponding trajectory $A_*(\cdot)$ is an *optimal admissible trajectory*.

3 Reduction to a One-dimensional Problem Without Integral Constraints

Let us introduce a new state variable $x(\cdot):[0,\infty)\to\mathbb{R}^1$ and new control variables $u(\cdot):[0,\infty)\to(0,\infty)$ and $v(\cdot):[0,\infty)\to[0,\infty)$ as follows:

$$x(t)=\frac{S(t)^{1-\eta}}{A(t)^{1-\theta}},\qquad u(t)=\frac{R_1(t)}{S(t)},\qquad v(t)=\frac{R_2(t)}{S(t)},\qquad t>0. \quad (12)$$

Here the state variable $S(\cdot)$ represents the current supply of the exhaustible resource. This variable is a (Carathéodory) solution to the following Cauchy problem (for given admissible controls $R_1(\cdot)$ and $R_2(\cdot)$) on $[0,\infty)$:

$$\dot S(t)=-R_1(t)-R_2(t),\qquad S(0)=S_0. \quad (13)$$

Note that the case $\eta=\theta=1$ is not excluded, although in this case the new variable $x(\cdot)$ degenerates into a constant. This case can easily be analyzed directly, but we include it in our general scheme to save the space. Below we show that for $\eta=\theta=1$ the problem reduces to a zero-dimensional problem, i.e. to a problem in which the utility function depends only on the controls and does not depend on the state variables (hence the control variables take constant values maximizing the utility function at each moment in time).

Note also that $S(t)>0$ for all $t>0$, so the quantities $u(t)$ and $v(t)$ are well defined for all $t>0$. Indeed, if $S(\tau)=0$ for some $\tau>0$, then $S(t)=0$ for all $t>\tau$

and hence $R_1(t) = R_2(t) = 0$ for $t > \tau$, which is precluded by (5). Moreover, $u(\cdot)$ and $v(\cdot)$ are locally bounded measurable functions since $R_i(\cdot)$, $i = 1, 2$, is locally bounded and measurable and $S(\cdot)$ is positive and continuous.

Since $x(\cdot)$ is a (locally) absolutely continuous function, we can calculate its derivative a.e. on $[0, \infty)$:

$$\begin{aligned}\dot{x}(t) &= (1-\eta)\frac{\dot{S}(t)}{A(t)^{1-\theta}S(t)^{\eta}} - (1-\theta)\frac{\dot{A}(t)S(t)^{1-\eta}}{A(t)^{2-\theta}}\\ &= -(1-\eta)\big[u(t)+v(t)\big]x(t) - (1-\theta)\frac{A(t)^{\theta}[L^A(t)]^{\eta}R_2(t)^{1-\eta}S(t)^{1-\eta}}{A(t)^{2-\theta}}\\ &= -(1-\eta)\big[u(t)+v(t)\big]x(t) - (1-\theta)\big[L^A(t)\big]^{\eta}v(t)^{1-\eta}x(t)^2.\end{aligned}$$

Thus, $x(\cdot)$ is a Carathéodory solution of the differential equation

$$\dot{x}(t) = -(1-\eta)\big[u(t)+v(t)\big]x(t) - (1-\theta)\big[L^A(t)\big]^{\eta}v(t)^{1-\eta}x(t)^2, \quad t > 0, \tag{14}$$

satisfying the initial condition

$$x(0) = x_0 = \frac{S_0^{1-\eta}}{A_0^{1-\theta}}. \tag{15}$$

Now we express the functional $J(A(\cdot), L^A(\cdot), R_1(\cdot))$ (see (7)) in terms of the new variables $x(\cdot)$, $u(\cdot)$ and $v(\cdot)$. Consider the first term in the integrand in (7):

$$\int_0^{\infty} e^{-\rho t}\ln A(t)dt = \frac{\ln A_0}{\rho} + \frac{1}{\rho}\int_0^{\infty} e^{-\rho t}\frac{\dot{A}(t)}{A(t)}dt. \tag{16}$$

This formula is valid for any admissible trajectory $A(\cdot)$ of problem (P). To show this, it suffices first to integrate by parts on a finite time interval $[0, T]$ and then pass to the limit as $T \to \infty$:

$$\int_0^{T} e^{-\rho t}\ln A(t)dt = \frac{\ln A_0 - e^{-\rho T}\ln A(T)}{\rho} + \frac{1}{\rho}\int_0^{T} e^{-\rho t}\frac{\dot{A}(t)}{A(t)}dt. \tag{17}$$

Due to (10) the integral on the left-hand side and the first term on the right-hand side tend to the corresponding terms in (16). Further, $\dot{A}(t) \geq 0$, $t > 0$; therefore, $e^{-\rho t}\dot{A}(t)/A(t)$ is integrable on $[0, +\infty)$ and the last term in (17) tends to the last term in (16).

Substituting $\dot{A}(t)$ from (4) into (16), we obtain

$$\begin{aligned}\int_0^{\infty} e^{-\rho t}\ln A(t)dt &= \frac{\ln A_0}{\rho} + \frac{1}{\rho}\int_0^{\infty} e^{-\rho t}\frac{A(t)^{\theta}[L^A(t)]^{\eta}v(t)^{1-\eta}S(t)^{1-\eta}}{A(t)}dt\\ &= \frac{\ln A_0}{\rho} + \frac{1}{\rho}\int_0^{\infty} e^{-\rho t}\big[L^A(t)\big]^{\eta}v(t)^{1-\eta}x(t)dt.\end{aligned}$$

Similarly,

$$
\begin{aligned}
&\int_0^T e^{-\rho t}\ln R_1(t)dt\\
&\quad=\int_0^T e^{-\rho t}\big[\ln u(t)+\ln S(t)\big]dt\\
&\quad=\int_0^T e^{-\rho t}\ln u(t)dt+\frac{\ln S_0-e^{-\rho T}\ln S(T)}{\rho}+\frac{1}{\rho}\int_0^T e^{-\rho t}\frac{\dot S(t)}{S(t)}dt\\
&\quad=\frac{\ln S_0-e^{-\rho T}\ln S(T)}{\rho}+\int_0^T e^{-\rho t}\bigg[\ln u(t)-\frac{u(t)+v(t)}{\rho}\bigg]dt.
\end{aligned}
$$

Passing to the limit as $T\to\infty$, we see that

$$\int_0^\infty e^{-\rho t}\ln R_1(t)dt=\frac{\ln S_0}{\rho}+\int_0^\infty e^{-\rho t}\bigg[\ln u(t)-\frac{u(t)+v(t)}{\rho}\bigg]dt,$$

where both sides may be $-\infty$.

Thus, multiplying $J(A(\cdot), L^A(\cdot), R_1(\cdot))$ by ρ and neglecting constant terms, we arrive at the functional

$$
\begin{aligned}
J_1\big(x(\cdot),L^A(\cdot),u(\cdot),v(\cdot)\big)=\int_0^\infty e^{-\rho t}\big\{\varkappa\big[L^A(t)\big]^\eta v(t)^{1-\eta}x(t)+\alpha\rho\ln\big[L-L^A(t)\big]\\
+(1-\alpha)\rho\ln u(t)-(1-\alpha)\big[u(t)+v(t)\big]\big\}dt. \qquad (18)
\end{aligned}
$$

Now consider the following optimal control problem (P1) (see (14), (15) and (18)):

$$\dot x(t)=-(1-\eta)\big[u(t)+v(t)\big]x(t)-(1-\theta)\big[L^A(t)\big]^\eta v(t)^{1-\eta}x(t)^2, \qquad (19)$$

$$v(t)\in[0,\infty),\qquad L^A(t)\in[0,L),\qquad u(t)\in(0,\infty), \qquad (20)$$

$$x(0)=x_0,$$

$$
\begin{aligned}
J_1\big(x(\cdot),L^A(\cdot),u(\cdot),v(\cdot)\big)=\int_0^\infty e^{-\rho t}\big\{\varkappa\big[L^A(t)\big]^\eta v(t)^{1-\eta}x(t)+\alpha\rho\ln\big[L-L^A(t)\big]\\
+(1-\alpha)\rho\ln u(t)-(1-\alpha)\big[u(t)+v(t)\big]\big\}dt\\
\to\max. \qquad (21)
\end{aligned}
$$

We say that a control $\tilde w(\cdot)=(L^A(\cdot),u(\cdot),v(\cdot)):[0,\infty)\to[0,L)\times(0,\infty)\times[0,\infty)$ (which is a triple of measurable functions) is admissible in problem (P1) if the functions $u(\cdot)$ and $v(\cdot)$ are locally bounded. The corresponding trajectory $x(\cdot):[0,\tau)\to\mathbb{R}^1$, $\tau>0$, can obviously be extended to the whole infinite time interval $[0,\infty)$. So, without loss of generality, we assume that any admissible trajectory $x(\cdot)$ is defined on $[0, \infty)$. A pair $(x(\cdot), w(\cdot))$ where $w(\cdot)$ is an admissible control and $x(\cdot)$ is the corresponding trajectory is called an admissible pair or a process in problem (P1).

Note that, structurally, problem (P1) is simpler than problem (P) because problem (P1) does not contain integral constraints on the control variables. Problem (P1) is equivalent to problem (P) in the following sense:

Lemma 1 *For fixed* A_0 *and* S_0, *there is a one-to-one correspondence between processes* $(A(\cdot), w(\cdot))$ *in problem* (P) *and* $(x(\cdot), \tilde{w}(\cdot))$ *in problem* (P1). *Moreover, the corresponding values of the objective functionals* $J(A(\cdot), L^A(\cdot), R_1(\cdot))$ *and* $J_1(x(\cdot), L^A(\cdot), u(\cdot), v(\cdot))$ *are related by a linear transformation of the form*

$$J_1\big(x(\cdot), L^A(\cdot), u(\cdot), v(\cdot)\big) = \rho J\big(A(\cdot), L^A(\cdot), R_1(\cdot)\big) + C, \tag{22}$$

where C *depends only on* ρ, A_0 *and* S_0.

Proof As shown above, any process $(A(\cdot), w(\cdot)) = (A(\cdot), L^A(\cdot), R_1(\cdot), R_2(\cdot))$ in problem (P) generates a process $(x(\cdot), \tilde{w}(\cdot)) = (x(\cdot), L^A(\cdot), u(\cdot), v(\cdot))$ in problem (P1), and relation (22) is valid for these processes.

Now, we show that any control process $(x(\cdot), \tilde{w}(\cdot)) = (x(\cdot), L^A(\cdot), u(\cdot), v(\cdot))$ in problem (P1) corresponds to a control process $(A(\cdot), w(\cdot)) = (A(\cdot), L^A(\cdot), R_1(\cdot), R_2(\cdot))$ in problem (P). First, using the controls $u(\cdot)$ and $v(\cdot)$, we determine $S(\cdot)$ as a unique solution to the Cauchy problem

$$\dot{S}(t) = -\big[u(t) + v(t)\big]S(t), \qquad S(0) = S_0.$$

Since $u(\cdot) + v(\cdot)$ is positive and locally bounded, we obtain a positive monotonically decreasing function $S(\cdot)$ defined on $[0, \infty)$. Then we define $R_1(t) = u(t)S(t)$ and $R_2(t) = v(t)S(t)$, $t \geq 0$, which are locally bounded and satisfy the integral constraint in (5). Finally, we find $A(\cdot)$ as a unique solution to the Cauchy problem

$$\frac{d}{dt}\big[A(t)^{1-\theta}\big] = (1-\theta)\big[L^A(t)\big]^{\eta} v(t)^{1-\eta} S(t)^{1-\eta}, \qquad A(0) = A_0$$

if $\theta < 1$, or as a unique solution to the Cauchy problem

$$\frac{d}{dt}\big[\ln A(t)\big] = \big[L^A(t)\big]^{\eta} v(t)^{1-\eta} S(t)^{1-\eta}, \qquad A(0) = A_0$$

if $\theta = 1$. This is certainly possible because the right-hand side of each of these equations is positive and locally bounded.

We thus have a process $(A(\cdot), w(\cdot)) = (A(\cdot), L^A(\cdot), R_1(\cdot), R_2(\cdot))$ in problem (P). Passing from this process $(A(\cdot), w(\cdot))$ in problem (P) back to some process $(x_1(\cdot), \tilde{w}_1(\cdot))$ in problem (P1) along the scheme described at the beginning of this section, we see that $\tilde{w}_1(\cdot) = \tilde{w}(\cdot)$ and $x_1(\cdot)$ satisfies the same Cauchy problem (14),

(15) as $x(\cdot)$. Therefore, by the uniqueness theorem for solutions of differential equations, $x_1(\cdot) = x(\cdot)$. This proves the required one-to-one correspondence between the admissible processes in problems (P) and (P1). Since (22) holds for the processes $(A(\cdot), w(\cdot))$ and $(x_1(\cdot), \tilde{w}_1(\cdot))$, and $(x_1(\cdot), \tilde{w}_1(\cdot)) = (x(\cdot), \tilde{w}(\cdot))$, we conclude that (22) is valid for $(A(\cdot), w(\cdot))$ and $(x(\cdot), \tilde{w}(\cdot))$. □

As a direct consequence of Lemma 1 and estimate (8) we arrive at

Lemma 2 *There exists a constant $M_1 > 0$ depending only on ρ, L, A_0 and S_0 such that*

$$\sup_{(x(\cdot),\tilde{w}(\cdot))} \int_0^\infty e^{-\rho t}\big\{\varkappa\big[L^A(t)\big]^\eta v(t)^{1-\eta} x(t) + \alpha\rho \ln\big[L - L^A(t)\big]$$
$$+ (1-\alpha)\rho \ln u(t) - (1-\alpha)\big[u(t) + v(t)\big]\big\}dt \le M_1,$$

where the supremum is taken over all admissible pairs $(x(\cdot), \tilde{w}(\cdot))$ in problem (P1).

Lemma 2 allows us to define an optimal control $\tilde{w}_*(\cdot) : [0, \infty) \to \mathbb{R}^3$ in problem (P1) as a welfare-maximizing triple $\tilde{w}_*(\cdot) = (L_*^A(\cdot), u_*(\cdot), v_*(\cdot))$. The corresponding admissible trajectory $x_*(\cdot)$ is an optimal one in problem (P1).

To recapitulate, we showed that a process $(A(\cdot), w(\cdot))$ is optimal in problem (P) if and only if the corresponding process $(x(\cdot), \tilde{w}(\cdot))$ is optimal in problem (P1). In the next section we formulate and prove two main theoretical results on which the subsequent solution of the problem is based.

4 Existence of an Optimal Control and Pontryagin's Maximum Principle

Denote

$$\begin{aligned} f(x,\ell,u,v) &= -(1-\eta)(u+v)x - (1-\theta)\ell^\eta v^{1-\eta} x^2, \\ g(x,\ell,u,v) &= \varkappa \ell^\eta v^{1-\eta} x + \alpha\rho \ln(L-\ell) + (1-\alpha)\rho \ln u \\ &\quad - (1-\alpha)(u+v), \\ x > 0, &\qquad \ell \in [0, L), \qquad u > 0, \qquad v \ge 0, \end{aligned} \tag{23}$$

so that (19) and (21) become

$$\dot{x}(t) = f\big(x(t), L^A(t), u(t), v(t)\big),$$
$$J_1\big(x(\cdot), L^A(\cdot), u(\cdot), v(\cdot)\big) = \int_0^\infty e^{-\rho t} g\big(x(t), L^A(t), u(t), v(t)\big)dt \to \max.$$

Let $\mathcal{M}(x, u, v, p)$ and $M(x, p)$ be the current value Hamilton–Pontryagin function and the current value Hamiltonian for problem (P1) in the normal form:

$$\begin{aligned}\mathcal{M}(x,\ell,u,v,p) &= f(x,\ell,u,v)p + g(x,\ell,u,v)\\ &= -(1-\eta)(u+v)xp - (1-\theta)\ell^{\eta}v^{1-\eta}x^2p + \varkappa\ell^{\eta}v^{1-\eta}x\\ &\quad + \alpha\rho\ln(L-\ell) + (1-\alpha)\rho\ln u - (1-\alpha)(u+v),\\ M(x,p) &= \sup_{\ell\in[0,L),u>0,v\geq 0}\mathcal{M}(x,\ell,u,v,p).\end{aligned} \tag{24}$$

Here $x>0$, $\ell\in[0,L)$, $u>0$, $v\geq 0$ and $p\in\mathbb{R}^1$.

Next, we formulate two important theorems (an existence theorem and a version of the Pontryagin maximum principle for problem (P1)) that allow us to perform a qualitative analysis of the solution to problem (P) (in Sect. 5). The proofs of these theorems (together with all necessary auxiliary statements) constitute the rest of this section.

Theorem 1 (Existence) *There exists an optimal process $(x_*(\cdot), \tilde{w}_*(\cdot))$ in problem* (P1). *The process $(A_*(\cdot), w_*(\cdot))$ corresponding to $(x_*(\cdot), \tilde{w}_*(\cdot))$ (in the sense of Lemma* 1) *is optimal in problem* (P).

Theorem 2 (Maximum principle) *Let $(x_*(\cdot), \tilde{w}_*(\cdot)) = (x_*(\cdot), L_*^A(\cdot), u_*(\cdot), v_*(\cdot))$ be an optimal process in problem* (P1) *and $(A_*(\cdot), w_*(\cdot))$ be the corresponding (in the sense of Lemma* 1) *optimal process in problem* (P). *Then there exists a current value adjoint variable $p(\cdot)$ such that the following conditions hold*:

(i) *The process $(x_*(\cdot), \tilde{w}_*(\cdot))$, together with the current value adjoint variable $p(\cdot)$, satisfies the core relations of the Pontryagin maximum principle in the normal form on the infinite time interval* $[0,\infty)$:

$$\dot{p}(t) = \rho p(t) - \frac{\partial\mathcal{M}(x_*(t), L_*^A(t), u_*(t), v_*(t), p(t))}{\partial x} \quad \textit{for a.e. } t>0, \tag{25}$$

$$\mathcal{M}\big(x_*(t), L_*^A(t), u_*(t), v_*(t), p(t)\big) = M\big(x_*(t), p(t)\big) \quad \textit{for a.e. } t>0. \tag{26}$$

(ii) *The process $(x_*(\cdot), \tilde{w}_*(\cdot))$, together with the current value adjoint variable $p(\cdot)$, satisfies the normal-form stationarity condition*

$$\begin{aligned} &M\big(x_*(t), p(t)\big)\\ &\quad = \rho e^{\rho t}\int_t^{\infty} e^{-\rho s} g\big(x_*(s), L_*^A(s), u_*(s), v_*(s)\big)ds \quad \textit{for all } t\geq 0.\end{aligned}$$

(iii) *For any $t\geq 0$*

$$p(t) = e^{\rho t}e^{-y(t)}\int_t^{\infty} e^{-\rho s}e^{y(s)}\frac{\partial g(x_*(s), L_*^A(s), u_*(s), v_*(s))}{\partial x}ds, \tag{27}$$

where $y(t) = \int_0^t \frac{\partial f(x_*(s), L_*^A(s), u_*(s), v_*(s))}{\partial x}ds \leq 0$.

Let us outline the scheme of proofs of these two theorems. First, we show that it suffices to consider only bounded controls in problem (P1). Then we introduce the problem with a slightly modified objective functional, which is defined for controls that take values in the compact closure of the admissible control set. We show that the optimal processes in these two problems coincide. Finally, using standard results of optimal control theory, we prove analogs of Theorems 1 and 2 for the modified problem, which automatically implies the assertions of Theorems 1 and 2. The above approach is presented as a series of auxiliary lemmas that are subsequently used to prove the theorems.

Denote

$$V_0 = \left(\frac{(1-\eta)\varkappa L^\eta x_0}{1-\alpha} \right)^{1/\eta} \tag{28}$$

and consider the following optimal control problem (P1′) with bounded controls:

$$\dot{x}(t) = -(1-\eta)\big[u(t)+v(t)\big]x(t) - (1-\theta)\big[L^A(t)\big]^\eta v(t)^{1-\eta} x(t)^2, \tag{29}$$

$$L^A(t) \in [0, L), \qquad u(t) \in (0, \rho], \qquad v(t) \in [0, V_0], \tag{30}$$

$$x(0) = x_0, \tag{31}$$

$$\begin{aligned} J_1\big(x(\cdot), L^A(\cdot), u(\cdot), v(\cdot)\big) = \int_0^\infty & e^{-\rho t}\big\{\varkappa\big[L^A(t)\big]^\eta v(t)^{1-\eta} x(t) + \alpha\rho \ln\big[L - L^A(t)\big] \\ & + (1-\alpha)\rho \ln u(t) - (1-\alpha)\big[u(t)+v(t)\big]\big\}dt \\ & \to \max. \end{aligned} \tag{32}$$

Lemma 3 *If* $\tilde{w}_*(\cdot) = (L_*^A(\cdot), u_*(\cdot), v_*(\cdot))$ *is an optimal admissible control in problem* (P1), *then*

$$u_*(t) \le \rho \quad \text{and} \quad v_*(t) \le V_0 = \left(\frac{(1-\eta)\varkappa L^\eta x_0}{1-\alpha} \right)^{1/\eta} \quad \text{for a.e. } t > 0,$$

and so $\tilde{w}_*(\cdot)$ *is also an optimal admissible control in problem* (P1′). *Conversely, if* $\hat{\tilde{w}}_*(\cdot)$ *is an optimal admissible control in problem* (P1′), *then it is also an optimal admissible control in problem* (P1).

Before proving the lemma, we point out a corollary to this lemma and formula (27).

Corollary 1 *The current value adjoint variable* $p(\cdot)$ *satisfying the conditions of Theorem* 2 *is bounded*:

$$0 \le p(t) \le \frac{\varkappa L^\eta V_0^{1-\eta}}{\rho} \quad \text{for all } t > 0$$

(*if* $\eta = 1$, *then* $V_0 = 0$ *and we consider* $V_0^{1-\eta}$ *to be* 1). *In particular, the transversality condition*

$$\lim_{t\to\infty} e^{-\rho t} x_*(t) p(t) = 0$$

holds for any optimal process $(x_*(\cdot), \tilde{w}_*(\cdot))$ *in problem* (P1).

Proof Indeed, since $\frac{\partial f}{\partial x}(x, \ell, u, v) \le 0$ for all $x > 0$, $\ell \in [0, L)$, $u > 0$ and $v \ge 0$, it follows that $y(\cdot)$ is a monotonically decreasing function, and so

$$0 \le p(t) \le e^{\rho t} \int_t^\infty e^{-\rho s} \varkappa L_*^A(s)^\eta v_*(s)^{1-\eta} ds \le \frac{\varkappa L^\eta V_0^{1-\eta}}{\rho} \quad \text{for all } t > 0.$$

This implies the transversality condition, as $0 < x_*(t) \le x_0$ for $t > 0$. □

Proof of Lemma 3 Let $\tilde{w}(\cdot) = (L^A(\cdot), u(\cdot), v(\cdot))$ be an admissible control in problem (P1) such that $\operatorname{ess\,sup}_{t>0} u(t) > \rho$ or $\operatorname{ess\,sup}_{t>0} v(t) > V_0$. Define a new admissible bounded control $\overline{w}(\cdot) = (L^A(\cdot), \bar{u}(\cdot), \bar{v}(\cdot))$ with $\bar{u}(t) = \min\{u(t), \rho\}$ and $\bar{v}(t) = \min\{v(t), V_0\}$, $t \ge 0$. Note that $\overline{w}(\cdot)$ is also an admissible control in problem (P1′).

Let $x(\cdot)$ and $\bar{x}(\cdot)$ be the trajectories of problem (P1) (with the same initial condition x_0) that correspond to $\tilde{w}(\cdot)$ and $\overline{w}(\cdot)$, respectively ($\bar{x}(\cdot)$ is also a trajectory of problem (P1′)). Then we have

$$\bar{u}(t) \le u(t), \qquad \bar{v}(t) \le v(t) \quad \text{and} \quad x_0 \ge \bar{x}(t) \ge x(t) > 0 \quad \text{for all } t > 0$$

by virtue of (19). Therefore,

$$\begin{aligned}
J_1\big(x(\cdot), L^A(\cdot), u(\cdot), v(\cdot)\big) &\le \int_0^\infty e^{-\rho t}\big\{\varkappa\big[L^A(t)\big]^\eta v(t)^{1-\eta}\bar{x}(t) + \alpha\rho\ln\big[L - L^A(t)\big] \\
&\quad + (1-\alpha)\rho\ln u(t) - (1-\alpha)\big[u(t) + v(t)\big]\big\}dt \\
&< \int_0^\infty e^{-\rho t}\big\{\varkappa\big[L^A(t)\big]^\eta \bar{v}(t)^{1-\eta}\bar{x}(t) + \alpha\rho\ln\big[L - L^A(t)\big] \\
&\quad + (1-\alpha)\rho\ln\bar{u}(t) - (1-\alpha)\big[\bar{u}(t) + \bar{v}(t)\big]\big\}dt \\
&= J_1\big(\bar{x}(\cdot), L^A(\cdot), \bar{u}(\cdot), \bar{v}(\cdot)\big),
\end{aligned}$$

where we used the inequalities

$$\frac{d}{du}\big((1-\alpha)\rho\ln u - (1-\alpha)u\big) < 0, \qquad \frac{d}{dv}\big(\varkappa\big[L^A(t)\big]^\eta v^{1-\eta}\bar{x}(t) - (1-\alpha)v\big) < 0$$

for all $t > 0$ and $u > \rho$, $v > V_0$.

Thus, we see that if $\operatorname{ess\,sup}_{t>0} u(t) > \rho$ or $\operatorname{ess\,sup}_{t>0} v(t) > V_0$, then the control $\tilde{w}(\cdot)$ cannot be optimal. This proves the first part of the lemma.

Conversely, if $(x_*(\cdot), \tilde{w}_*(\cdot)) = (x_*(\cdot), L^A_*(\cdot), u_*(\cdot), v_*(\cdot))$ is an optimal process in problem (P1′) and $(x(\cdot), \tilde{w}(\cdot)) = (x(\cdot), L^A(\cdot), u(\cdot), v(\cdot))$ is any process in problem (P1), then, again, introducing a new bounded control $\overline{w}(\cdot) = (L^A(\cdot), \bar{u}(\cdot), \bar{v}(\cdot))$ with $\bar{u}(t) = \min\{u(t), \rho\}$ and $\bar{v}(t) = \min\{v(t), V_0\}$, $t \geq 0$, we see that

$$J_1\big(x(\cdot), L^A(\cdot), u(\cdot), v(\cdot)\big) \leq J_1\big(\bar{x}(\cdot), L^A(\cdot), \bar{u}(\cdot), \bar{v}(\cdot)\big)$$
$$\leq J_1\big(x_*(\cdot), L^A_*(\cdot), u_*(\cdot), v_*(\cdot)\big),$$

where $\bar{x}(\cdot)$ is the trajectory of problem (P1) (as well as of (P1′)) corresponding to the control $\overline{w}(\cdot)$. □

Our next goal is to establish the existence of an optimal admissible control $\tilde{w}_*(\cdot)$ in problem (P1′). To apply a standard existence theorem of optimal control theory, we need to compactify the range of values of the control variables. For this purpose, we introduce the function

$$\mathcal{L}_\varepsilon(\xi) = \begin{cases} \ln \varepsilon + \dfrac{1}{\varepsilon}(\xi - \varepsilon) & \text{for } 0 \leq \xi \leq \varepsilon, \\ \ln \xi & \text{for } \xi > \varepsilon, \end{cases} \tag{33}$$

where $\varepsilon < 1$ is a small positive constant, to the utility functional $J_1(x(\cdot), L^A(\cdot), u(\cdot), v(\cdot))$. Obviously, $\mathcal{L}_\varepsilon(\cdot)$ is a continuously differentiable and concave function on $[0, \infty)$ and $\mathcal{L}_\varepsilon(\xi) \geq \ln \xi$ for $\xi \in (0, \infty)$.

Now consider an auxiliary problem (P_ε):

$$\dot{x}(t) = -(1-\eta)\big[u(t) + v(t)\big]x(t) - (1-\theta)\big[L^A(t)\big]^\eta v(t)^{1-\eta} x(t)^2, \tag{34}$$

$$L^A(t) \in [0, L], \qquad u(t) \in [0, \rho], \qquad v(t) \in [0, V_0], \tag{35}$$

$$x(0) = x_0,$$

$$J_\varepsilon\big(x(\cdot), L^A(\cdot), u(\cdot), v(\cdot)\big) = \int_0^\infty e^{-\rho t}\big\{\varkappa\big[L^A(t)\big]^\eta v(t)^{1-\eta} x(t) + \alpha\rho\mathcal{L}_\varepsilon\big(L - L^A(t)\big)$$
$$+ (1-\alpha)\rho\mathcal{L}_\varepsilon\big(u(t)\big) - (1-\alpha)\big[u(t) + v(t)\big]\big\}dt$$
$$\to \max, \tag{36}$$

where x_0 is the same as in (31). Clearly, any process $(x(\cdot), \tilde{w}(\cdot)) = (x(\cdot), L^A(\cdot), u(\cdot), v(\cdot))$ in problem (P1′) is also an admissible process in problem (P_ε).

Lemma 4 *If there is an optimal process* $(x_*(\cdot), \tilde{w}_*(\cdot)) = (x_*(\cdot), L^A_*(\cdot), u_*(\cdot), v_*(\cdot))$ *in problem* (P_ε) *such that* $L^A_*(t) \leq L - \varepsilon$ *and* $u_*(t) \geq \varepsilon$ *for a.e.* $t \in (0, \infty)$, *then*

(i) *this process is also optimal in problem* (P1′);

(ii) *any other optimal process* $(\hat{x}_*(\cdot), \hat{\tilde{w}}_*(\cdot)) = (\hat{x}_*(\cdot), \hat{L}_*^A(\cdot), \hat{u}_*(\cdot), \hat{v}_*(\cdot))$ *(if it exists) in problem* (P1′) *is such that* $\hat{L}_*^A(t) \le L - \varepsilon$ *and* $\hat{u}_*(t) \ge \varepsilon$ *for a.e.* $t \in (0, \infty)$ *and so it is also optimal in problem* (P_ε).

Proof Assertion (i) is valid because $J_\varepsilon(x(\cdot), L^A(\cdot), u(\cdot), v(\cdot)) \ge J_1(x(\cdot), L^A(\cdot), u(\cdot), v(\cdot))$ for any admissible process $(x(\cdot), \tilde{w}(\cdot)) = (x(\cdot), L^A(\cdot), u(\cdot), v(\cdot))$ in problem (P1′), while $J_\varepsilon(x_*(\cdot), L_*^A(\cdot), u_*(\cdot), v_*(\cdot)) = J_1(x_*(\cdot), L_*^A(\cdot), u_*(\cdot), v_*(\cdot))$.

If $(\hat{x}(\cdot), \hat{\tilde{w}}(\cdot)) = (\hat{x}(\cdot), \hat{L}^A(\cdot), \hat{u}(\cdot), \hat{v}(\cdot))$ is a process in problem (P1′) such that $\hat{L}^A(t) > L - \varepsilon$ or $\hat{u}(t) < \varepsilon$ on a positive measure set of values of t, then

$$
\begin{aligned}
J_1\big(\hat{x}(\cdot), \hat{L}^A(\cdot), \hat{u}(\cdot), \hat{v}(\cdot)\big) &< J_\varepsilon\big(\hat{x}(\cdot), \hat{L}^A(\cdot), \hat{u}(\cdot), \hat{v}(\cdot)\big) \\
&\le J_\varepsilon\big(x_*(\cdot), L_*^A(\cdot), u_*(\cdot), v_*(\cdot)\big) \\
&= J_1\big(x_*(\cdot), L_*^A(\cdot), u_*(\cdot), v_*(\cdot)\big)
\end{aligned}
$$

and hence this process cannot be optimal in problem (P1′). This implies (ii). □

Denote

$$
W = [0, L] \times [0, \rho] \times [0, V_0]
$$

and

$$
\begin{aligned}
g_\varepsilon(x, \ell, u, v) = {}& \varkappa \ell^\eta v^{1-\eta} x + \alpha\rho \mathcal{L}_\varepsilon(L - \ell) \\
&+ (1-\alpha)\rho\mathcal{L}_\varepsilon(u) - (1-\alpha)(u+v), \\
x > 0, \qquad & (\ell, u, v) \in W,
\end{aligned}
\tag{37}
$$

so that (34) and (36) become

$$
\dot{x}(t) = f\big(x(t), L^A(t), u(t), v(t)\big),
$$

$$
J_\varepsilon\big(x(\cdot), L^A(\cdot), u(\cdot), v(\cdot)\big) = \int_0^\infty e^{-\rho t} g_\varepsilon\big(x(t), L^A(t), u(t), v(t)\big) dt \to \max
$$

(see (23)).

For every $x > 0$, consider the following set, which is standard in optimal control theory:

$$
Q(x) = \big\{(z^0, z) \in \mathbb{R}^2 : z^0 \le g_\varepsilon(x, \ell, u, v), z = f(x, \ell, u, v), (\ell, u, v) \in W\big\}.
$$

Lemma 5 *For every* $x > 0$, *the set* $Q(x)$ *is convex.*

Proof It suffices to show that for any two points $(z_1^0, z_1), (z_2^0, z_2) \in Q(x)$ the midpoint of the segment joining (z_1^0, z_1) to (z_2^0, z_2) also lies in $Q(x)$. Let $z_i = f(x, \ell_i, u_i, v_i)$ and $z_i^0 \le g_\varepsilon(x, \ell_i, u_i, v_i)$ for some $(\ell_i, u_i, v_i) \in W$ $(i = 1, 2)$. We need to show that there exists $(\bar{\ell}, \bar{u}, \bar{v}) \in W$ such that

$$f(x, \bar{\ell}, \bar{u}, \bar{v}) = \bar{z} = \frac{z_1 + z_2}{2} \quad \text{and} \quad g_\varepsilon(x, \bar{\ell}, \bar{u}, \bar{v}) \geq \bar{z}^0 = \frac{z_1^0 + z_2^0}{2}.$$

We will seek $(\bar{\ell}, \bar{u}, \bar{v})$ in the form

$$\bar{\ell} = \bar{\ell}(\epsilon) = \frac{\ell_1 + \ell_2}{2} - \epsilon, \qquad \bar{u} = \frac{u_1 + u_2}{2}, \qquad \bar{v} = \frac{v_1 + v_2}{2}$$

with $0 \leq \epsilon \leq \frac{\ell_1 + \ell_2}{2}$. It is obvious that such a triple belongs to W.

Note that

$$\left(\frac{\ell_1 + \ell_2}{2}\right)^\eta \left(\frac{v_1 + v_2}{2}\right)^{1-\eta} \geq \frac{\ell_1^\eta v_1^{1-\eta} + \ell_2^\eta v_2^{1-\eta}}{2}, \qquad 0 \leq \eta \leq 1$$

(see, e.g., Theorem 38 in Hardy et al. 1934). Therefore,

$$f(x, 0, \bar{u}, \bar{v}) \geq \bar{z} \quad \text{and} \quad f\big(x, \bar{\ell}(0), \bar{u}, \bar{v}\big) \leq \bar{z}.$$

Since $f(x, \bar{\ell}(\cdot), \bar{u}, \bar{v})$ is a continuous function of ϵ, there indeed exists an ϵ, $0 \leq \epsilon \leq \frac{\ell_1 + \ell_2}{2}$, such that

$$f\big(x, \bar{\ell}(\epsilon), \bar{u}, \bar{v}\big) = \bar{z}. \tag{38}$$

We fix such an ϵ and write simply $\bar{\ell}$ instead of $\bar{\ell}(\epsilon)$ in what follows.

Now let us show that $g_\varepsilon(x, \bar{\ell}, \bar{u}, \bar{v}) \geq \bar{z}^0$. Note that due to (38), for $\theta < 1$,

$$\begin{aligned} \bar{\ell}^\eta \bar{v}^{1-\eta} x &= \frac{-(1-\eta)(\bar{u} + \bar{v})x - \bar{z}}{(1-\theta)x} = \frac{-(1-\eta)(u_1 + u_2 + v_1 + v_2)x - (z_1 + z_2)}{2(1-\theta)x} \\ &= \frac{\ell_1^\eta v_1^{1-\eta} x + \ell_2^\eta v_2^{1-\eta} x}{2}. \end{aligned} \tag{39}$$

If $\theta = 1$, then $f(\cdot)$ does not depend on ℓ and so (38) holds for all ϵ. Therefore, choosing an appropriate ϵ, we can achieve the equality of the first and last expressions in the chain (39) in this case as well.

Since $\mathcal{L}_\varepsilon(\cdot)$ is a concave increasing function, we have $\mathcal{L}_\varepsilon(L - \bar{\ell}) \geq \mathcal{L}_\varepsilon(L - \bar{\ell}(0))$ and in view of (39) find that

$$g_\varepsilon(x, \bar{\ell}, \bar{u}, \bar{v}) \geq \frac{g_\varepsilon(x, \ell_1, u_1, v_1) + g_\varepsilon(x, \ell_2, u_2, v_2)}{2} \geq \bar{z}^0.$$

This completes the proof of Lemma 5. □

Lemma 6 *For any ε, $0 < \varepsilon < 1$, there exists an optimal control in problem* (P_ε). *Moreover, if ε is small enough, then any optimal control $\tilde{w}(\cdot) = (L_*^A(\cdot), u_*(\cdot), v_*(\cdot))$ in problem (P_ε) is such that $L_*^A(t) \leq L - \varepsilon$ and $u_*(t) \geq \varepsilon$ for a.e. $t \in (0, \infty)$.*

Proof The existence follows from Theorem 2.1 in Aseev and Kryazhimskii (2007) and Lemma 5.

Note that problem (P_ε) falls within the case of dominating discount (see Sect. 12 in Aseev and Kryazhimskii 2007), so we can apply the version of Pontryagin's maximum principle formulated in Theorem 12.1 in Aseev and Kryazhimskii (2007) to this problem. To this end, define the current value Hamilton–Pontryagin function $\mathcal{M}_\varepsilon(x,u,v,p)$ and the current value Hamiltonian $M_\varepsilon(x,p)$ in problem (P_ε) in the normal form:

$$\begin{aligned}\mathcal{M}_\varepsilon(x,\ell,u,v,p) &= f(x,\ell,u,v)p + g_\varepsilon(x,\ell,u,v)\\ &= -(1-\eta)(u+v)xp - (1-\theta)\ell^\eta v^{1-\eta}x^2 p + \varkappa \ell^\eta v^{1-\eta} x\\ &\quad + \alpha\rho\mathcal{L}_\varepsilon(L-\ell) + (1-\alpha)\rho\mathcal{L}_\varepsilon(u) - (1-\alpha)(u+v), \quad (40)\\ M_\varepsilon(x,p) &= \sup_{(\ell,u,v)\in W} \mathcal{M}_\varepsilon(x,\ell,u,v,p). \quad (41)\end{aligned}$$

Here $x>0$, $(\ell,u,v)\in W$ and $p\in\mathbb{R}^1$.

Let $(x_*(\cdot),\tilde{w}_*(\cdot)) = (x_*(\cdot), L_*^A(\cdot), u_*(\cdot), v_*(\cdot))$ be an optimal process in problem (P_ε). Then, by Theorem 12.1 from Aseev and Kryazhimskii (2007), we have

$$\mathcal{M}_\varepsilon\big(x_*(t), L_*^A(t), u_*(t), v_*(t), p(t)\big) = M_\varepsilon\big(x_*(t), p(t)\big) \quad \text{for a.e. } t>0, \qquad (42)$$

where

$$p(t) = e^{\rho t}e^{-y(t)}\int_t^\infty e^{-\rho s}e^{y(s)}\frac{\partial g_\varepsilon(x_*(s), L_*^A(s), u_*(s), v_*(s))}{\partial x}ds \qquad (43)$$

with the same $y(\cdot)$ as in Theorem 2. As shown in the proof of Corollary 1, $y(\cdot)$ is a monotonically decreasing function, and so

$$0 \le p(t) \le \frac{1}{\rho}\sup_{x>0,(\ell,u,v)\in W}\frac{\partial g_\varepsilon(x,\ell,u,v)}{\partial x} = \frac{\varkappa L^\eta V_0^{1-\eta}}{\rho} \quad \text{for all } t>0.$$

We also have $0 < x_*(\cdot) \le x_0$. However, it is easy to show that if ε is sufficiently small,[1] then the maximum of the function $\mathcal{M}_\varepsilon(x,\cdot,\cdot,\cdot,p)$ with respect to $(\ell,u,v)\in W$ for fixed $x\in(0,x_0]$ and $p\in[0,\varkappa L^\eta V_0^{1-\eta}/\rho]$ cannot be attained at a point (ℓ,u,v) such that $\ell > L-\varepsilon$ or $u<\varepsilon$. Indeed, it suffices to calculate the partial derivatives of $\mathcal{M}_\varepsilon$ with respect to ℓ and u.

This fact, together with the maximum condition (42), completes the proof of the lemma. □

Proof of Theorem 1 Above we have shown that the auxiliary problem (P_ε) has a solution, i.e. an optimal process $(x_*(\cdot),\tilde{w}_*(\cdot)) = (x_*(\cdot), L_*^A(\cdot), u_*(\cdot), v_*(\cdot))$, and

[1] Of course, the upper bound for ε that guarantees the validity of this statement depends on x_0, but x_0 is fixed from the onset.

proved certain estimates for the corresponding optimal control (Lemma 6). These estimates show (Lemma 4) that any such solution is also an optimal process in problem (P1′), and so is an optimal process in problem (P1) (Lemma 3), which is equivalent to the original problem (P) (Lemma 1). Thus, we obtain the existence of an optimal control in problem (P). □

Proof of Theorem 2 Fix a sufficiently small ε. By Lemmas 6 and 4(ii), $L_*^A(t) \le L - \varepsilon$ and $u_*(t) \ge \varepsilon$ for a.e. $t \in (0, \infty)$, and $(x_*(\cdot), \tilde{w}_*(\cdot))$ is an optimal process in problem (P_ε).

By Theorem 12.1 in Aseev and Kryazhimskii (2007), such an adjoint variable $p(\cdot)$ satisfying properties (i)–(iii) of Theorem 2 (with $g_\varepsilon(\cdot)$, $M_\varepsilon(\cdot)$ and $\mathcal{M}_\varepsilon(\cdot)$ instead of $g(\cdot)$, $M(\cdot)$ and $\mathcal{M}(\cdot)$, respectively) exists for the optimal process $(x_*(\cdot), \tilde{w}_*(\cdot))$ in problem (P_ε). Since $L_*^A(t) \le L - \varepsilon$ and $u_*(t) \ge \varepsilon$ for a.e. $t > 0$, we have $g(x_*(t), L_*^A(t), u_*(t), v_*(t)) = g_\varepsilon(x_*(t), L_*^A(t), u_*(t), v_*(t))$ and $\mathcal{M}(x_*(t), L_*^A(t), u_*(t), v_*(t), p(t)) = \mathcal{M}_\varepsilon(x_*(t), L_*^A(t), u_*(t), v_*(t), p(t))$ for a.e. $t > 0$. Moreover, since $\mathcal{M}(x, \ell, u, v, p) \le \mathcal{M}_\varepsilon(x, \ell, u, v, p)$ for all $x > 0$, $p > 0$ and $(\ell, u, v) \in W$, we also have $M(x_*(t), p(t)) = M_\varepsilon(x_*(t), p(t))$.

Thus, properties (i)–(iii) of Theorem 2 with $g(\cdot)$, $M(\cdot)$ and $\mathcal{M}(\cdot)$ follow from the same properties with $g_\varepsilon(\cdot)$, $M_\varepsilon(\cdot)$ and $\mathcal{M}_\varepsilon(\cdot)$. In particular, (42) and (43) become (26) and (27). □

Theorem 2 allows us to explicitly write the Hamiltonian system of the Pontryagin maximum principle for problem (P1). In the next section, we will analyze the qualitative behavior of solutions to this system and single out all optimal regimes.

5 Analysis of the Hamiltonian System

We know from Theorem 1 that an optimal process $(x_*(\cdot), \tilde{w}_*(\cdot))$ in problem (P1) exists and satisfies the relations of Theorem 2. Using Theorem 2, we can construct the Hamiltonian system of the Pontryagin maximum principle for problem (P1) in the variables $x(\cdot)$ and $p(\cdot)$ directly. However, to simplify the further analysis, we pass from the variable $p(\cdot)$ to a new variable $\phi(\cdot)$ defined as $\phi(t) = x(t)p(t)$, $t > 0$. Then we write and analyze the relations of the Hamiltonian system of the Pontryagin maximum principle for problem (P1) in the variables $x(\cdot)$ and $\phi(\cdot)$.

In terms of the variable $\phi(\cdot)$, the adjoint system (see (25)) and the maximum condition (see (26)) take the forms

$$\begin{aligned}\dot{\phi}(t) &= \dot{x}(t)p(t) + x(t)\dot{p}(t)\\ &= \rho\phi(t) + L^A(t)^\eta v(t)^{1-\eta} x(t)\big[(1-\theta)\phi(t) - \varkappa\big]\end{aligned} \tag{44}$$

and

$$\tilde{\mathcal{M}}(x, \ell, u, v, \phi) \to \max_{\ell \in [0, L), u > 0, v \ge 0}, \tag{45}$$

respectively. Here the function $\tilde{\mathcal{M}}(\cdot)$ is defined by the equality (see (24))

$$\begin{aligned}\tilde{\mathcal{M}}(x,\ell,u,v,\phi) = &-\big[1-\alpha+(1-\eta)\phi\big](u+v)\\ &+\big[\varkappa-(1-\theta)\phi\big]\ell^{\eta}v^{1-\eta}x\\ &+\alpha\rho\ln(L-\ell)+(1-\alpha)\rho\ln u,\end{aligned} \tag{46}$$

for all $x>0$, $\phi\geq 0$, $u>0$, $v\geq 0$ and $0\leq \ell < L$.

Our first aim is to write the Hamiltonian system of the maximum principle for problem (P1) in terms of the variables $x(\cdot)$ and $\phi(\cdot)$ by combining (19) and (44) (and using maximum condition (45)). To this end, we first express the quantities $L^A(x,\phi)$, $u(x,\phi)$ and $v(x,\phi)$ as functions of x and ϕ that are (unique) maximizers of $\tilde{\mathcal{M}}(\cdot)$ with respect to ℓ, u and v, respectively (see maximum condition (45)), for all $x>0$ and $\phi\geq 0$. Then, substituting these maximizers into (19) and (44), we get the Hamiltonian system of the maximum principle for problem (P1) in the form

$$\begin{aligned}\dot{x}(t) &= -(1-\eta)\big[u\big(x(t),\phi(t)\big)+v\big(x(t),\phi(t)\big)\big]x(t)\\ &\quad -(1-\theta)L^A\big(x(t),\phi(t)\big)^{\eta}v\big(x(t),\phi(t)\big)^{1-\eta}x(t)^2,\\ \dot{\phi}(t) &= \rho\phi(t)+L^A\big(x(t),\phi(t)\big)^{\eta}v\big(x(t),\phi(t)\big)^{1-\eta}x(t)\big[(1-\theta)\phi(t)-\varkappa\big].\end{aligned} \tag{47}$$

The value $u(x,\phi)$ at which the maximum of $\tilde{\mathcal{M}}(\cdot)$ with respect to u is attained can easily be found by means of differentiation (see (46)):

$$u(x,\phi) = \frac{(1-\alpha)\rho}{1-\alpha+(1-\eta)\phi}. \tag{48}$$

If $\varkappa \leq (1-\theta)\phi$, then the maximum of $\tilde{\mathcal{M}}(\cdot)$ with respect to ℓ and v is attained for $v(x,\phi)=L^A(x,\phi)=0$.

Suppose that $\varkappa > (1-\theta)\phi$. If $\eta=1$, then $v(x,\phi)=0$ simply because of the constraint $0\leq v\leq V_0=0$ (see (35) and (28)), and $u(x,\phi)=\rho$ (see (48)). In this case it is obvious that the maximum point of $\tilde{\mathcal{M}}(\cdot)$ as a function of ℓ is given by

$$L^A(x,\phi) = L - \frac{\alpha\rho}{(\varkappa-(1-\theta)\phi)x}. \tag{49}$$

Finally, consider the case when $\varkappa>(1-\theta)\phi$ and $\eta<1$. Note that $\tilde{\mathcal{M}}(x,\ell,u,v,\phi)\to-\infty$ as $v\to\infty$ or $\ell\to L-0$. On the other hand, if one of the variables, v or ℓ, is zero, then the maximum with respect to the other variable is attained at zero. Therefore, the maximum of $\tilde{\mathcal{M}}(\cdot)$ with respect to ℓ and v is attained either at the point $v(x,\phi)=L^A(x,\phi)=0$ or at an interior point, in which case this point can be found by equating the partial derivatives of $\tilde{\mathcal{M}}(\cdot)$ with respect to ℓ and v to zero:

$$\eta\big[\varkappa-(1-\theta)\phi\big]\left(\frac{v}{\ell}\right)^{1-\eta}x = \frac{\alpha\rho}{L-\ell}, \tag{50}$$

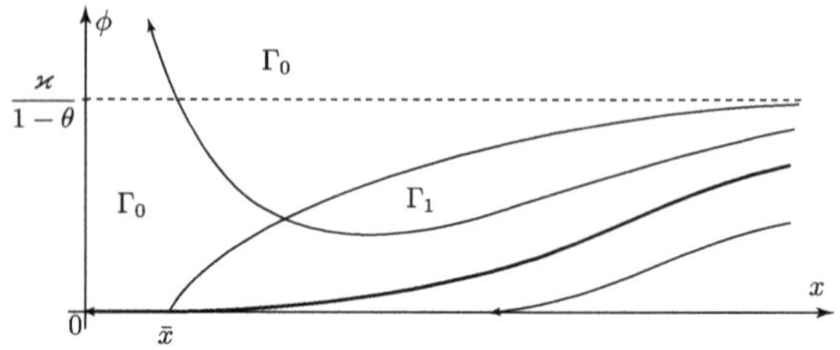

Fig. 1 The sets Γ_0 and Γ_1 and the optimal trajectory (*thick line*). All trajectories lying above the optimal one tend to infinity along the ϕ-*axis*. All trajectories lying below the optimal one transversally intersect the x-*axis*

$$(1-\eta)\big[\varkappa-(1-\theta)\phi\big]\left(\frac{\ell}{v}\right)^{\eta}x = 1-\alpha+(1-\eta)\phi. \tag{51}$$

Denoting

$$h(x,\phi) = \frac{1-\alpha+(1-\eta)\phi}{(1-\eta)x[\varkappa-(1-\theta)\phi]}, \qquad x>0, 0\le\phi<\frac{\varkappa}{1-\theta},$$

we find

$$\frac{\ell}{v} = h(x,\phi)^{1/\eta} \tag{52}$$

and

$$\ell = L - \frac{\alpha\rho h(x,\phi)^{(1-\eta)/\eta}}{\eta[\varkappa-(1-\theta)\phi]x} = L - \frac{\alpha\rho(1-\alpha+(1-\eta)\phi)^{(1-\eta)/\eta}}{\eta(1-\eta)^{(1-\eta)/\eta}(x[\varkappa-(1-\theta)\phi])^{1/\eta}}, \tag{53}$$

$$\begin{aligned} v &= \frac{L}{h(x,\phi)^{1/\eta}} - \frac{\alpha\rho h(x,\phi)^{-1}}{\eta[\varkappa-(1-\theta)\phi]x} \\ &= \frac{L((1-\eta)x[\varkappa-(1-\theta)\phi])^{1/\eta}}{(1-\alpha+(1-\eta)\phi)^{1/\eta}} - \frac{\alpha\rho(1-\eta)}{\eta(1-\alpha+(1-\eta)\phi)}. \end{aligned} \tag{54}$$

If these formulas yield positive values $v(x,\phi)$ and $L^A(x,\phi)$ of v and ℓ, then this is the maximum point of $\tilde{\mathcal{M}}(\cdot)$ with respect to v and ℓ. Otherwise, the maximum point is $v(x,\phi) = L^A(x,\phi) = 0$.

Note that (53) and (54) for $\eta = 1$ turn into (49) and $v(x,\phi) = 0$, respectively, if we consider $(1-\eta)^{1-\eta}$ to be 1 for $\eta = 1$.

Set

$$h_1(\phi) = \frac{\alpha^{\eta}\rho^{\eta}(1-\alpha+(1-\eta)\phi)^{1-\eta}}{L^{\eta}\eta^{\eta}(1-\eta)^{1-\eta}[\varkappa-(1-\theta)\phi]}, \qquad 0\le(1-\theta)\phi<\varkappa,$$

and introduce the following sets (see Fig. 1):

$$\Gamma = \big\{(x,\phi)\in\mathbb{R}^2 : x>0, \phi\ge 0\big\},$$

$$\Gamma_0 = \big\{(x,\phi)\in\Gamma : (1-\theta)\phi\ge\varkappa \text{ or } \big\{(1-\theta)\phi<\varkappa, x<h_1(\phi)\big\}\big\}, \qquad \Gamma_1 = \Gamma\setminus\Gamma_0.$$

According to the above analysis, in Γ_0 both $L^A(x,\phi)$ and $v(x,\phi)$ vanish, and so our Hamiltonian system (47) in Γ_0 has the form

$$\dot{x}(t) = -\frac{(1-\eta)(1-\alpha)\rho}{1-\alpha+(1-\eta)\phi(t)}x(t),$$
$$\dot{\phi}(t) = \rho\phi(t).$$

Note that $h_1(\cdot)$ is a monotonically increasing function of ϕ (except for the case $\eta = \theta = 1$, in which $h_1(\cdot) \equiv \text{const}$). Therefore, any trajectory of our system that reaches the set Γ_0 cannot leave this set afterward. (Indeed, at every point of Γ_0 we have $\dot{x}(\cdot) \le 0$ and $\dot{\phi}(\cdot) \ge 0$.) However, we know that $\phi(\cdot)$ is bounded along an optimal trajectory (e.g., by Corollary 1); hence the only candidate for an optimal trajectory in Γ_0 lies on the x-axis and looks like

$$x(t) = \bar{x}e^{-(1-\eta)\rho(t-\tau)}, \qquad \phi(t) = 0 \quad \text{for } t \ge \tau, \tag{55}$$

where

$$\bar{x} = h_1(0) = \frac{\rho^\eta\alpha^\eta(1-\alpha)^{1-\eta}}{L^\eta\eta^\eta(1-\eta)^{1-\eta}\varkappa}. \tag{56}$$

On the other hand, since $\dot{x}(t) \le 0$, any bounded trajectory must tend to a fixed point. If $\eta < 1$, then $\dot{x}(\cdot) < 0$ in the interior of Γ_1 and consequently any trajectory of our system starting in Γ_1 eventually enters the set Γ_0. This shows that there is a unique bounded trajectory of our system, and hence the optimal process in problem (P1) is also unique. The tail of this trajectory is described by (55).

If $\eta = 1$ and $\theta < 1$, then for similar reasons any bounded trajectory starting in Γ_1 tends to the point $(\bar{x}, 0)$ on the boundary of Γ_1. Let us show that there is only one such trajectory $(\tilde{x}(\cdot), \tilde{\phi}(\cdot))$ in Γ_1. Indeed, if there were two trajectories lying in Γ_1 and tending to $(\bar{x}, 0)$, then any trajectory lying between these two would also tend to $(\bar{x}, 0)$ (because $\dot{x}(\cdot) \le 0$). However, this is impossible, as we can show, for example, by considering the linearization of the Hamiltonian system of the maximum principle in Γ_1 at the point $(\bar{x}, 0)$ and applying the Grobman–Hartman theorem (see Hartman 1964).

Finally, if $\eta = \theta = 1$, then $x(t) \equiv 1$ (see (12)) and $\dot{\phi}(t) = \rho\phi(t) - \ell\varkappa$, where $\ell = \max\{0, L - \frac{\alpha\rho}{\varkappa}\}$. Thus, the only bounded trajectory is the fixed point $x = 1$, $\phi = \max\{0, \frac{L\varkappa}{\rho} - \alpha\}$. Recall that in this case the optimal controls are $u(t) \equiv \rho$, $v(t) \equiv 0$ and $L^A(t) \equiv \max\{0, L - \frac{\alpha\rho}{\varkappa}\}$.

Let us now examine the initial part of the optimal trajectory lying in Γ_1, for $\eta < 1$. Using formulas (52) and (54), we find

$$\ell^\eta v^{1-\eta} = h(x,\phi)v = \frac{L}{h(x,\phi)^{(1-\eta)/\eta}} - \frac{\alpha\rho}{\eta x[\varkappa - (1-\theta)\phi]}.$$

Similarly, due to (48) and (54), we obtain

$$u + v = \frac{(\eta-\alpha)\rho}{\eta(1-\alpha+(1-\eta)\phi)} + \frac{L}{h(x,\phi)^{1/\eta}}.$$

Thus, our system (47) in Γ_1 has the form

$$\begin{aligned}\dot{x}(t) &= -(1-\eta)\left[\frac{(\eta-\alpha)\rho}{\eta(1-\alpha+(1-\eta)\phi(t))} + \frac{L}{h(x(t),\phi(t))^{1/\eta}}\right]x(t)\\ &\quad -(1-\theta)\left[\frac{L}{h(x(t),\phi(t))^{(1-\eta)/\eta}} - \frac{\alpha\rho}{\eta x(t)[\varkappa-(1-\theta)\phi(t)]}\right]x(t)^2, \qquad (57)\\ \dot{\phi}(t) &= \rho\phi(t) - \frac{L(1-\alpha+(1-\eta)\phi(t))}{(1-\eta)h(x(t),\phi(t))^{1/\eta}} + \frac{\alpha\rho}{\eta},\end{aligned}$$

and we are interested in the trajectory $(\tilde{x}(\cdot),\tilde{\phi}(\cdot))$ that passes through the point $(\bar{x},0)$. It would be difficult to solve this system analytically, but for numerical simulations it suffices to know that the sought trajectory $(\tilde{x}(\cdot),\tilde{\phi}(\cdot))$ is a solution to the Cauchy problem for system (57) in reverse time (i.e., with the right-hand side taken with the opposite sign) under the initial condition $\tilde{x}(0)=\bar{x}$, $\tilde{\phi}(0)=0$.

Moreover, since $\dot{\tilde{x}}(t)<0$ for all $t>0$, we can express $\tilde{\phi}(\cdot)$ as a function of $\tilde{x}(\cdot)$ along this trajectory, $\tilde{\phi}=\phi_*(x)$.

If $\eta=1$ and $\theta<1$, we can also express $\tilde{\phi}(\cdot)$ as a (continuous) function of $\tilde{x}(\cdot)$ along this trajectory, $\tilde{\phi}=\phi_*(x)$ (with $\phi_*(x)=0$ for $x\le\bar{x}$). However, this trajectory cannot be found as a solution of the Cauchy problem, as described above, as $(\bar{x},0)$ is a fixed point of the Hamiltonian system for $\eta=1$.

Thus, for $\eta\theta<1$ we obtain a unique optimal feedback control $u_*(x)=u(x,\phi_*(x))$, $v_*(x)=v(x,\phi_*(x))$, $L_*^A(x)=L^A(x,\phi_*(x))$ according to formulas (48), (54) and (49), (53).

Let us summarize the above analysis of the Hamiltonian system as follows:

Theorem 3

(a) *If* $\eta=1$ *and* $\theta=1$, *then there is a unique optimal control* $\tilde{w}(\cdot)=(L_*^A(\cdot),u_*(\cdot),v_*(\cdot))$ *in problem* (P1), *with*

$$L_*^A(t)\equiv\max\left\{0,L-\frac{\alpha\rho}{\varkappa}\right\},\qquad u_*(t)\equiv\rho,\qquad v_*(t)\equiv 0\quad \textit{for all } t\in[0,\infty).$$

In this case $x(t)\equiv x_0=1$, $t\ge 0$ *is a unique admissible trajectory* (*see* (12)).

(b) *If* $\eta\theta<1$, *then there is a unique optimal feedback control* (*optimal synthesis*) $\tilde{w}_*(x)=(L_*^A(x),u_*(x),v_*(x))$ *in problem* (P1), *with* $L_*^A(x)=L^A(x,\phi_*(x))$, $u_*(x)=u(x,\phi_*(x))$ *and* $v_*(x)=v(x,\phi_*(x))$ *determined by formulas* (49), (53), (48) *and* (54). *Here the feedback* $\phi_*(x)$ *is generated by a unique solution* $(\tilde{x}(\cdot),\tilde{\phi}(\cdot))$ *of the Hamiltonian system* (57) *that reaches* (*or tends to*) *the point* $(\bar{x},0)$ *from the right, where* (*see* (56))

$$\bar{x}=\frac{\rho^\eta\alpha^\eta(1-\alpha)^{1-\eta}}{L^\eta\eta^\eta(1-\eta)^{1-\eta}\varkappa}.$$

Namely,

(b.1) *If* $\eta\theta < 1$ *and* $x \le \bar{x}$, *then*

$$L_*^A(x) = 0, \qquad u_*(x) = \rho, \qquad v_*(x) = 0.$$

(b.2) *If* $\eta = 1$, $\theta < 1$ *and* $x > \bar{x}$, *then (see* (49), (48) *and* (54))

$$L_*^A(x) = L - \frac{\alpha\rho}{(\varkappa - (1-\theta)\phi_*(x))x}, \qquad u_*(x) = \rho, \qquad v_*(x) = 0.$$

In the case of $\eta = 1$ *and* $\theta < 1$, *for any initial state* $x_0 \le \bar{x}$ *the corresponding optimal trajectory* $x_*(\cdot)$ *is* $x_*(t) \equiv x_0$, $t \ge 0$, *while for any initial state* $x_0 > \bar{x}$ *the corresponding optimal trajectory* $x_*(\cdot)$ *monotonically tends to the point* $\bar{x}$ *from the right as* $t \to \infty$.

(b.3) *If* $\eta < 1$, $\theta \le 1$ *and* $x > \bar{x}$, *then (see* (53), (48) *and* (54))

$$L_*^A(x) = L - \frac{\alpha\rho(1-\alpha+(1-\eta)\phi_*(x))^{(1-\eta)/\eta}}{\eta(1-\eta)^{(1-\eta)/\eta}(x[\varkappa-(1-\theta)\phi_*(x)])^{1/\eta}},$$

$$u_*(x) = \frac{(1-\alpha)\rho}{1-\alpha+(1-\eta)\phi_*(x)},$$

$$v_*(x) = \frac{L((1-\eta)x[\varkappa-(1-\theta)\phi_*(x)])^{1/\eta}}{(1-\alpha+(1-\eta)\phi_*(x))^{1/\eta}} - \frac{\alpha\rho(1-\eta)}{\eta(1-\alpha+(1-\eta)\phi_*(x))}.$$

In the case of $\eta < 1$ *and* $\theta \le 1$, *for any initial state* $x_0 > 0$, *the corresponding optimal trajectory* $x_*(\cdot)$ *monotonically decreases to* 0 *as* $t \to \infty$.

Finally let us analyze the dynamics of the output $Y(\cdot)$ and the knowledge stock $A(\cdot)$ along the optimal trajectory.

If $\eta = \theta = 1$, then (Theorem 3(a)) the optimal controls are $u(t) \equiv \rho$, $v(t) \equiv 0$ and $L^A(t) \equiv \max\{0, L - \frac{\alpha\rho}{\varkappa}\}$. In the case of $L\varkappa \le \alpha\rho$, we have stagnation of the knowledge stock ($\dot{A}(t) \equiv 0$) and depletion of the output ($Y(t) \to 0$ as $t \to \infty$). For $L\varkappa > \alpha\rho$, the knowledge stock grows exponentially, while the output still depletes to zero for $L\varkappa < \rho(\alpha + \varkappa(1-\alpha))$, is constant for $L\varkappa = \rho(\alpha + \varkappa(1-\alpha))$, and grows exponentially for $L\varkappa > \rho(\alpha + \varkappa(1-\alpha))$.

Let us consider the case $\eta\theta < 1$ in more detail. If $x_0 \le \bar{x}$, then we again have stagnation of the knowledge stock and depletion of the output. If $x_0 > \bar{x}$, then the knowledge stock grows in the beginning, but the growth either terminates at a certain instant ($\eta < 1$) or decelerates ($\eta = 1$), so that the knowledge stock never exceeds a certain level determined by the parameters of the system. The output falls to zero in the long run. However, the following proposition shows that it may grow on some initial time interval.

Theorem 4 *Let* $\eta\theta < 1$. *Then, for sufficiently large initial values* x_0 *(i.e., for a relatively large initial stock of the exhaustible resource* S_0 *and/or for a relatively small initial knowledge stock* A_0; *see* (15)), *the output* $Y(\cdot)$ *as a function of* t *increases on some initial time interval* $0 < t < \tau$, $\tau > 0$.

Proof For large x_0 the initial part of the optimal trajectory lies in Γ_1 and hence $Y(\cdot)$ is continuously differentiable for the corresponding values of t. Let us show that $\dot{Y}(t) > 0$ on the initial time interval $0 < t < \tau$, $\tau > 0$, of the optimal trajectory. We have

$$\begin{aligned}\dot{Y}(t) &= Y(t)\left[\varkappa\frac{\dot{A}(t)}{A(t)} - \alpha\frac{\dot{L}^A(t)}{L - L^A(t)} + (1-\alpha)\frac{\dot{u}(t)}{u(t)} + (1-\alpha)\frac{\dot{S}(t)}{S(t)}\right] \\ &= Y(t)\left[\varkappa L^A(t)^\eta v(t)^{1-\eta}x(t) - \alpha\frac{\dot{L}^A(t)}{L - L^A(t)}\right. \\ &\quad \left. + (1-\alpha)\frac{\dot{u}(t)}{u(t)} - (1-\alpha)\big(u(t) + v(t)\big)\right] \end{aligned} \tag{58}$$

(see (1), (2), (13) and (12)), where $u(t) = u_*(x(t))$, $v(t) = v_*(x(t))$ and $L^A(t) = L_*^A(x(t))$.

Let us show that $\dot{\phi}(t) < 0$ along the optimal trajectory in Γ_1. To see this, note that the curve on which $\dot{\phi}(t) = 0$ in Γ_1 is described by the equation

$$\rho\phi + \frac{\alpha\rho}{\eta} = \frac{L(1-\alpha+(1-\eta)\phi)}{(1-\eta)h(x,\phi)^{1/\eta}} = \frac{L(1-\eta)^{(1-\eta)/\eta}(x[\varkappa - (1-\theta)\phi])^{1/\eta}}{(1-\alpha+(1-\eta)\phi)^{(1-\eta)/\eta}}. \tag{59}$$

This equation defines x as a monotonically increasing function of ϕ. So any trajectory of our system that intersects this curve at some instant τ (at a point different from $(\bar{x}, 0)$) acquires a positive derivative of the ϕ-coordinate and later enters the set Γ_0 (at a point different from $(\bar{x}, \phi)$). Such a trajectory tends to infinity and so it is not optimal. Hence our optimal trajectory lies in Γ_1 completely below the above curve, and $\dot{\phi}(t) < 0$ on it. This immediately implies that $\dot{u}(t) \geq 0$ in (58) (see (48)).

To estimate the second term in the square brackets in (58), we first denote $\zeta(t) = x(t)[\varkappa - (1-\theta)\phi(t)]$, $\zeta_*(x) = x[\varkappa - (1-\theta)\phi_*(x)]$, and calculate (along the optimal trajectory in Γ_1)

$$\begin{aligned}\dot{\zeta}(t) &= \frac{d}{dt}\big(x(t)\big[\varkappa - (1-\theta)\phi(t)\big]\big) = \dot{x}(t)\big[\varkappa - (1-\theta)\phi(t)\big] - (1-\theta)x(t)\dot{\phi}(t) \\ &= -(1-\eta)\big[u_*\big(x(t)\big) + v_*\big(x(t)\big)\big]\zeta(t) - (1-\theta)\rho x(t)\phi(t) < 0, \end{aligned} \tag{60}$$

because $\zeta(t) > 0$ for $(x(t), \phi(t)) \in \Gamma_1$. Then, after some calculations, we find from (49) for $\eta = 1$, from (53) for $\eta < 1$, and from (44), (60) that

$$\begin{aligned} -\frac{\dot{L}^A(t)}{L - L^A(t)} &= \frac{(1-\eta)^2\dot{\phi}(t)}{\eta(1-\alpha+(1-\eta)\phi(t))} - \frac{1}{\eta}\frac{\dot{\zeta}(t)}{\zeta(t)} \\ &> -\frac{(1-\eta)^2 L^A(t)^\eta v(t)^{1-\eta}\zeta(t)}{\eta(1-\alpha+(1-\eta)\phi(t))} + \frac{(1-\theta)\rho x(t)\phi(t)}{\zeta(t)}. \end{aligned} \tag{61}$$

If $\eta = 1$ and $\theta < 1$, then the right-hand side of (61) is positive; hence $\frac{dL_*^A(x)}{dx} > 0$ and

$$L_*^A(x)^\eta v_*(x)^{1-\eta} x \to +\infty \quad \text{as } x \to +\infty. \tag{62}$$

This obviously implies that $\dot{Y}(t) > 0$ for large $x(t)$ along the optimal trajectory, as the second and third terms in the square brackets in (58) are nonnegative, while the last term is bounded due to the restrictions $u(t) \le \rho$ and $v(t) = 0$.

If $\eta < 1$ and $\theta \le 1$, then $\phi_*(x) < \varkappa/(1-\theta)$ in Γ_1. Let us show that $\phi_*(x) \to \varkappa/(1-\theta)$ as $x \to \infty$. Indeed, suppose the contrary. Then it follows from (53) that $L_*^A(x) \to L$ as $x \to \infty$, and due to (52) $v_*(x) \sim x^{1/\eta}$ as $x \to \infty$. Therefore,

$$\begin{aligned}\frac{d\phi_*(x)}{dx} &= \frac{\dot{\phi}(t)}{\dot{x}(t)} \\ &= \frac{L_*^A(x)^\eta v_*(x)^{1-\eta} x[\varkappa - (1-\theta)\phi_*(x)] - \rho\phi_*(x)}{(1-\eta)[u_*(x) + v_*(x)]x + (1-\theta)L_*^A(x)^\eta v_*(x)^{1-\eta} x^2} \sim \frac{1}{x},\end{aligned} \tag{63}$$

which contradicts the boundedness of $\phi_*(\cdot)$. Thus, $\phi_*(x) \to \varkappa/(1-\theta)$ as $x \to \infty$.

If $\zeta_*(\cdot)$ is unbounded, then by (53) $L_*^A(x) \to L$ as $x \to \infty$, and by (52) $v_*(x) \sim \zeta_*(x)^{1/\eta} = o(x^{1/\eta})$ and $v_*(x) \to \infty$ as $x \to \infty$. This shows that the first term in the square brackets in (61) dominates all the negative terms there, and so $\dot{Y}(t) > 0$ for large $x(t)$ along the optimal trajectory.

If $\zeta_*(\cdot)$ is bounded, then $v_*(\cdot)$ is bounded by (51). Hence the right-hand side of (61) is positive for large $x(t)$ and, in particular, $\frac{dL_*^A(x)}{dx} > 0$ for large x. Therefore, again by (51), $v_*(x)$ is bounded away from zero for large x. We see that (62) holds in this case as well, which again implies that $\dot{Y}(t) > 0$ for large $x(t)$ along the optimal trajectory.

Finally, consider the case of $\eta < 1$ and $\theta = 1$. In this case $\zeta(t) = \varkappa x(t)$. Multiplying (50) raised to the power η by (51) raised to the power $1-\eta$, we find that

$$\eta^\eta (1-\eta)^{1-\eta} \varkappa x = \frac{\alpha^\eta \rho^\eta (1 - \alpha + (1-\eta)\phi_*(x))^{1-\eta}}{(L - L_*^A(x))^\eta}$$

Recall that $\phi_*(\cdot)$ is a monotonically increasing function of x. If it were bounded, then we would have $L_*^A(x) \to L$ as $x \to \infty$, $v_*(x)^\eta \sim x$ by (51), and hence (63) would be valid, which is impossible for a bounded $\phi_*(\cdot)$. Thus, $\phi_*(x) \to \infty$ as $x \to \infty$.

On the other hand, $\phi_*(x) = O(x)$ because the optimal trajectory lies below the curve described by (59). Therefore, $L_*^A(x) \to L$ as $x \to \infty$ by (53) and $v_*(x) \ge v_0$ for some $v_0 > 0$ and for all sufficiently large x by (54). At the same time, $v_*(x)^\eta = o(x)$ by (54). This shows that the first term in the square brackets in (61) dominates all the negative terms there, and so $\dot{Y}(t) > 0$ for large $x(t)$ along the optimal trajectory.

We showed that for $\eta\theta < 1$ the output $Y(t)$ increases on some initial time interval provided that the initial supply of exhaustible resource S_0 is large and/or the initial knowledge stock A_0 is small. □

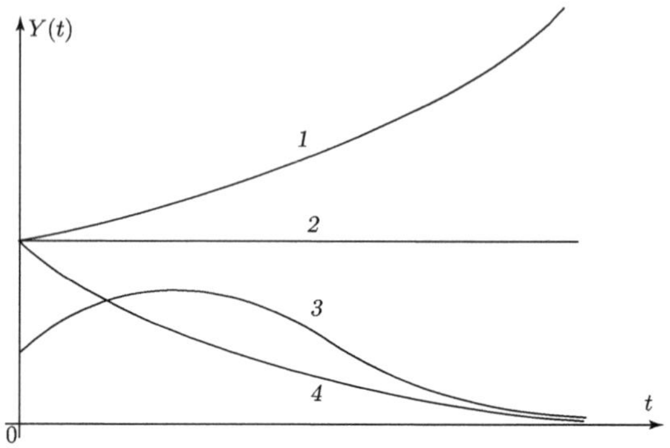

Fig. 2 Dynamics of the output $Y(\cdot)$ under optimal resource allocation: (*1*) $\eta = \theta = 1$, $L\varkappa > \rho(\alpha + \varkappa(1-\alpha))$; (2) $\eta = \theta = 1$, $L\varkappa = \rho(\alpha + \varkappa(1-\alpha))$; (*3*) $\eta\theta < 1$; (*4*) $\eta = \theta = 1$, $L\varkappa < \rho(\alpha + \varkappa(1-\alpha))$

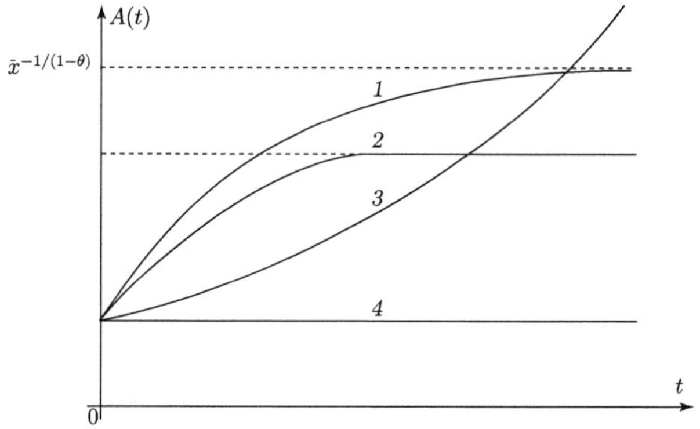

Fig. 3 Dynamics of the knowledge stock $A(\cdot)$ under optimal resource allocation: (*1*) $\eta = 1, \theta < 1$; (*2*) $\eta < 1$; (*3*) $\eta = \theta = 1$, $L\varkappa > \alpha\rho$; (*4*) $\eta = \theta = 1$, $L\varkappa \leq \alpha\rho$

6 Discussion

Dynamics of the output $Y(\cdot)$ and the knowledge stock $A(\cdot)$ along the optimal trajectory are depicted in Figs. 2 and 3. It follows from the above analysis that optimal growth is only sustainable if the following three conditions hold simultaneously:

(i) the exhaustible resource is not an input to the production of knowledge;
(ii) the accumulation of knowledge has strong scale effects;
(iii) the population is not too small.

In this scenario the growth of output is exponential. The resulting dynamics correspond to the optimal balanced growth path. In this case for a sufficiently large population size L, a constant fraction of labor is allocated to research. The lower the discount rate ρ, the higher this fraction. The fraction also depends on the elasticity of substitution in the production function. The optimal extraction policy implies an exponential depletion of the stock of the exhaustible resource, with the rate equal to the discount rate. This is the well-known Hotelling rule for the optimal depletion of exhaustible resources (see Hotelling 1931). In sum this implies an exponential growth of the knowledge stock $A(\cdot)$. Note that unlike in Jones (1995, 1999) and many other models, this balanced growth is fully endogenous in the sense of not requiring an exogenous population growth.

Requirement of strong scale effects (ii) for a balanced growth is the opposite to that obtained by Jones (2004). In his model, strong scale effects, coupled with an exponential growth of labor supply, lead to a double-exponential growth of output. This result follows from the assumption of an exponential population growth, which is unrealistic in the long run (Weeks 2004). Exponential population growth implies constant birth and death rates that are independent of the current population density. Second, more relevant here, exponential growth implies an arbitrarily large population in the long run. This is problematic in view of a finite resource base, a defining feature of our framework.

In the most realistic case $\eta\theta < 1$ we may have two qualitatively different optimal policies depending on whether the accumulation of knowledge requires the resource:

(i) When the accumulation of knowledge is independent of the resource ($\eta = 1$), the fraction of labor employed in research tends from an initially positive value to zero. This means that the research effort becomes successively smaller. The extraction policy is identical to that in the case of optimal sustainable growth described above. The stock of the exhaustible resource depletes exponentially with the rate equal to the discount rate (the Hotelling rule). The policy described above is optimal provided the initial knowledge stock is not too large ($x_0 > \bar{x}$). Otherwise it is optimal to allocate the entire labor to production from the onset.

(ii) When the accumulation of knowledge requires the resource ($\eta < 1$), it is optimal to conduct research until a certain ratio (characterized by (56)) between the knowledge stock and the current supply of the resource is reached. In this case the labor and resource allocated to research gradually decrease and ultimately vanish at the moment of reaching the above-mentioned ratio. Afterward the research effort stops and the stock of knowledge remains at its maximum level. This policy is optimal when $x_0 > \bar{x}$. For $x_0 \leq \bar{x}$ it is optimal not to invest in research as the initial knowledge stock is sufficiently large; the optimal extraction policy follows the Hotelling rule in this case.

Finally, condition (iii) says that a sufficiently small economy (with $L\varkappa \leq \alpha\rho$) will not grow, even under strong scale effects and even if the accumulation of knowledge does not depend on the exhaustible resource. This minimum size condition is the least restrictive of all conditions and can be assumed to hold a priori. In the typical case $\varkappa = 1$, we have $L > \alpha\rho$. This inequality can be maintained in all cases of interest since L is the size of the labor force, $\alpha < 1$ and ρ is the discount rate. The case $L\varkappa \leq \alpha\rho$ is included for completeness.

Acknowledgements The research was supported by the Austrian National Bank (OeNB) (Jubiläumsfonds project no. 13585) and the Russian Foundation for Basic Research (projects nos. 10-01-91004-ASF-a and 11-01-12112-ofi-m-2011). The authors would like to thank Arkady Kryazhimskiy, Tapio Palokangas, Gerald Silverberg, and the participants of the Green Growth and Sustainable Development symposium at IIASA for valuable comments and suggestions.

References

Acemoglu, D. (2008). *Introduction to modern economic growth*. Princeton: Princeton University Press.

Arrow, K. J., & Kurz, M. (1970). *Public investment, the rate of return, and optimal fiscal policy*. Baltimore: Johns Hopkins Press.

Aseev, S. M., & Kryazhimskii, A. V. (2007). The Pontryagin maximum principle and optimal economic growth problems. *Proceedings of the Steklov Institute of Mathematics*, *257*, 1–255.

Aseev, S. M., & Kryazhimskiy, A. V. (2004). The Pontryagin maximum principle and transversality conditions for a class of optimal control problems with infinite time horizons. *SIAM Journal of Control and Optimization*, *43*, 1094–1119.

Aubin, J.-P., & Clarke, F. H. (1979). Shadow prices and duality for a class of optimal control problems. *SIAM Journal of Control and Optimization*, *17*, 567–586.

Benveniste, L. M., & Scheinkman, J. A. (1982). Duality theory for dynamic optimization models of economics: the continuous time case. *Journal of Economic Theory*, *27*, 1–19.

Cabo, F., Martín-Herrán, G., & Martínez-Garćia, M. P. (2010). *Exhaustible resources and fully-endogenous growth with non-scale effects* (Les Cahiers du GERAD 33).

Cesari, L. (1983). *Optimization—theory and applications: problems with ordinary differential equations*. New York: Springer.

Dasgupta, P., & Heal, G. M. (1974). The optimal depletion of exhaustible resources. *Review of Economic Studies*, *41*, 3–28.

Filippov, A. F. (1988). *Differential equations with discontinuous right-hand sides*. Dordrecht: Kluwer.

Groth, C. (2006). *A new-growth perspective on non-renewable resources* (Discussion Paper 06-26). Department of Economics, University of Copenhagen.

Halkin, H. (1974). Necessary conditions for optimal control problems with infinite horizons. *Econometrica*, *42*, 267–272.

Hardy, G. H., Littlewood, J. E., & Pólya, G. (1934). *Inequalities*. Cambridge: Cambridge University Press.

Hartman, P. (1964). *Ordinary differential equations*. New York: Wiley.

Hotelling, H. (1931). The economics of exhaustible resources. *Journal of Political Economy*, *39*, 137–175.

Jones, C. I. (1995). Time-series test of endogenous growth models. *Quarterly Journal of Economics*, *110*, 495–525.

Jones, C. I. (1999). Growth: with or without scale effects. *American Economic Review*, *89*, 139–144.

Jones, C. I. (2004). *Growth and ideas*. Department of Economics, U.C.Berkley and NBER.

Pontrjagin, L. S., Boltyanskii, V. G., Gamkrelidze, R. V., & Mishchenko, E. F. (1964). *The mathematical theory of optimal processes*. Oxford: Pergamon.

Romer, P. (1986). Cake eating, chattering, and jumps: existence results for variational problems. *Econometrica*, *54*(4), 897–908.

Segerstrom, P. S. (1998). Endogenous growth without scale effects. *American Economic Review*, *88*, 1290–1310.

Seierstad, A., & Sydsæter, K. (1987). *Optimal control theory with economic applications*. Amsterdam: North-Holland.

Shell, K. (1969). Applications of Pontryagin's maximum principle to economics. In *Lecture notes in operation research mathematical economics: Vol. 11. Mathematical systems theory and economics 1* (pp. 241–292). Berlin: Springer.

Stiglitz, J. (1974). Growth with exhaustible natural resources: efficient and optimal growth paths. *Review of Economic Studies*, *41*, 123–137.

Weeks, J. R. (2004). In *Population: an introduction to concepts and issues*. Belmont: Wadsworth.

Weitzman, M. L. (2003). *Income, wealth, and the maximum principle*. Cambridge: Harvard University Press.

Young, A. (1998). Growth without scale effects. *Journal of Political Economy*, *106*, 41–63.

Optimal Pollution, Optimal Population, and Sustainability

Ulla Lehmijoki

1 Introduction

Is it possible that current utility maximization takes place at the cost of human lives? This possibility was already implied in the long-run consumer optimization models of Dasgupta and Heal (1974), Solow (1974), Stiglitz (1974), Krautkraemer (1985), and Pezzey and Withagen (1998) who argued that *the scarcity of natural resources* may lead to ever-decreasing per capita consumption. Per capita consumption may also decrease if *excessive pollution* impairs production and compromises life-supporting systems as was argued by Keeler et al. (1971), Plourde (1972), Foster (1973) and Smulders and Gradus (1996). However, in both types of models the demographic aspect is deficient as population either keeps constant or grows at a constant rate in spite of decreasing consumption numbers.

In this paper, I explicitly assume that population is endogenous to the environment, i.e., there is feedback from the environment to mortality which rises if population is not environmentally supported, this feedback being defined as a "positive check" by Robert Malthus (1914). Positive check may occur either because of the increasing scarcity of resources or because of the continuing concentration of pollutants. In this paper, I focus on pollutants as emerging evidence on the lethal effects of the pollutants maintains that the positive check is at work. This evidence consists of medical and econometric studies, showing that there already is a statistically significant increase in mortality due to urban air pollution, and that climate change may induce further increases in the future. Other global concerns, such as the pollution of ground waters and oceans, are also possible, but less evidence on their mortality effects has been received thus far.[1]

[1] In spite of my emphasis on pollutants, the model can be generalized to natural resources since resource depletion can be seen as pollution in the extended sense (Keeler et al. 1971).

U. Lehmijoki (✉)
University of Helsinki and HECER, "Economicum" Building, Arkadiankatu 7, P.O. Box 17, 00014 Helsinki, Finland
e-mail: Ulla.Lehmijoki@helsinki.fi

J. Crespo Cuaresma et al. (eds.), *Green Growth and Sustainable Development*, Dynamic Modeling and Econometrics in Economics and Finance 14,
DOI 10.1007/978-3-642-34354-4_2,

Section 2 of this paper reviews the empirical evidence on the positive check and Sect. 3 introduces a model of optimal pollution with endogenous population. Section 4 discusses the sustainability implications and a new definition for sustainability is supplied. The role of technical progress is shown to be less positive than what is usually suggested. Section 5 gives a parametric example and Sect. 6 closes the paper. To concentrate on population, the simplest model of optimal pollution is provided. Even so, endogenous population tends to make the model "murky" (Solow 1974) but excessive complexity can be avoided by modeling in virtual time.

2 The Positive Check—Recent Evidence

This Section reviews the global evidence on environmental mortality with focus on air pollution and climate change (CO_2 emissions). Mortality induced by air pollution has been debated since the smog in the Meuse Valley in 1930 and London in 1952 took the lives of 60 and 4000 people (Nemery et al. 2001 and Logan 1953). Recently, the Clean Air for Europe program (*CAFE*) and *WHO* have summarized the European research by collecting 629 peer-reviewed time-series studies and 160 individual or panel studies up to February 2003 (WHO 2004). In the original studies, daily adult mortality in several European cities was regressed against daily changes in air pollution as indicated by particular matter (PM) and ozone.[2] The summary estimates show that there is a statistically significant 0.6% and 0.3% increase in mortality for each 10 μg/m^3 increase in *PM* and ozone respectively.

The study for the effects of long-term *PM* exposure got its onset in the United States as Pope et al. (2002) analyzed questionnaires from 1982 which provided data on sex, race, smoking, alcohol consumption, etc., so that controlling for alternative risk sources was possible. The mortality data which were collected until 1998 implied that there was 4%, 6%, and 8% increases in all-cause, respiratory, and lung cancer mortality respectively for each 10 μg/m^3 increase in *PM*. Evans and Smith have estimated similar increases (Evans and Smith 2005). For a recent review of long-term study literature, see Raaschou-Nielsen et al. (2011). The estimates of Pope et al. (2002) were applied to the European data by *CAFE* and *WHO* to calculate that the short-term and long-term exposures were together responsible for 370 000 premature deaths in 2000 in Europe (WHO 2004). The infant mortality risk

[2]Air pollution consists of several components, of which particulate matter (*PM*) and ozone are the most dangerous (WHO 2004). The term particulate matter (*PM*) refers to solid airborne particles of varying size, chemical composition and origin. For example, the particles in PM_{10} have a diameter of less than 10 μm and are mainly combustion-derived, either from traffic or from energy production, often from long-distance sources. Existing evidence suggests that the smaller the particles are, the more deeply into the lung they penetrate (WHO 2004). Air pollution increases mortality mainly through an increase in respiratory and cardiovascular diseases and lung cancer (Samet et al. 2000), but an increase in skin cancer is also reported (Brunekreef and Holgate 2002). All age groups are affected, but unborn and young children as well as the elderly are the most vulnerable (Pope and Dockery 2006).

has been studied by for example by Currie and Neidell (2004), Chay and Greenstone (2003) and Scheers et al. (2011). WHO has summarized that, taken all types of deaths together, urban outdoor air pollution causes 1.3 million deaths worldwide per year (WHO 2011).

In climate-change studies, the mortality estimates are based on simulations (Pitcher et al. 2008). Tanser et al. (2003), for example, have applied the Hadley Centre's climate model to estimate that the increase in malaria distribution and the prolonged malaria season would lead to a 25% increase in the risk of death from malaria by 2100, mainly in Africa. The abundant literature on climate change has been collected and analyzed by the UN's Intergovernmental Panel on Climate Change (*IPCC*). Its Third Assessment Report suggests that mortality will increase because of weather extremes, because of environmental changes which lead to diseases or to water and food shortages, or because of conflicts in displaced populations (IPCC 2003, updated 2007). Relying on the *IPCC*, *WHO* has published a summary report on human health and climate change (WHO 2003). This report projects a maximum increase in the risk of 83%, 17%, and 32% for the great killers; malaria, diarrhoea, and malnutrition, respectively. There is also a great projected risk increase in coastal floods, but the number of deaths may be low (Gosling et al. 2009). The mortality effects of climate change are unequally distributed and are particularly severe in countries with already high disease burdens, such as sub-Saharan Africa and Asia (IPCC 2003). Nevertheless, Deschênes and Greenstone (2011) suggest that, under a business-as-usual scenario, climate change will also lead to an increase in the overall U.S. annual mortality rate ranging from 0.5% to 1.7% by the end of the 21st century. WHO has also summarized that, currently, climate change contributes to 150 000 deaths each year (WHO 2012).

3 The Model

3.1 Modeling the Positive Check

To model the positive check, note that the population growth rate $\dot{L}/L = n$ is the difference between fertility and mortality. In what follows, I assume that only mortality depends on pollution while fertility is constant.[3]

Pollution may increase mortality (decrease population growth) both as emissions E and as stocks S, but it seems appropriate to model in terms of stocks because their mortality effects are more longstanding. Hence, let:

$$n = n(S), \qquad n(0) > 0, \qquad n'(S) < 0, \qquad n(\hat{S}) = 0, \tag{1}$$

[3]Some studies suggest, however, that fertility may respond to environmental degradation both because it is causing poverty and because toxins etc. cause miscarriage (Lutz et al. 2005). Because the emphasis of this paper is on the positive check, the fertility effects are excluded, for simplicity.

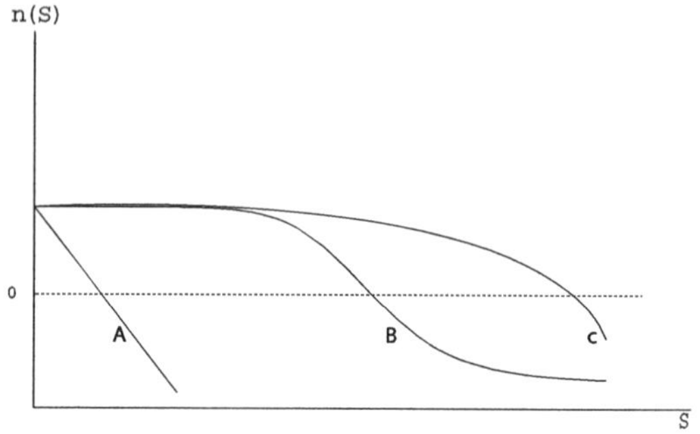

Fig. 1 Possible functional formulas for the positive check. Meadows et al. (1972)

where $\hat{S}$ is the critical stock beyond which population starts to decrease. Normalizing the initial level of population to unity it holds

$$L(t) = \exp \int_0^t n\big[S(\tau)\big] d\tau. \tag{2}$$

Several functional formulas satisfy the assumption of the population growth function (1). Some alternatives, repeated in Fig. 1, have been suggested already in the Report of Rome (Meadows et al. 1972). In A population growth decreases linearly, in C the negative effect is exponential, and in B mortality increases as pollution stock bypasses the threshold level after which the positive check cuts in and mortality starts to increase (population growth starts to decrease). Section 5 gives a closer look at these alternative cases.

The accumulation of the pollution stock is dictated by emissions and abatements which are given by an abatement function $\delta(S)$,

$$\dot{S} = E - \delta(S). \tag{3}$$

The first component of (3) can be rewritten as $E = (E/L) \cdot L$ to see that the environmental burden of population comes from two sources, namely from per capita emission E/L and from the number of people L.

The role of the second component, the abatement function $\delta(S)$ has been broadly debated in the literature.[4] In this paper, I assume a simple hump-shaped abatement function which is strictly concave. Thus, let $\delta(0) = \delta(\tilde{S}) = 0$ and $\delta'(0) > 0$, $\delta'(\tilde{S}) < 0$, $\delta''(S) < 0$ where $\tilde{S} > 0$ is the carrying capacity of the environment. To allow the possibility of negative population growth in the area $0 < S < \tilde{S}$, I assume $\hat{S} < \tilde{S}$.

3.2 The Household Optimization

Consider an infinitely living representative household which wants to maximize its Benthamian total utility. At each instant of time, the total utility then becomes

[4]For a review, see Tahvonen and Salo (1996).

$u(C/L) \cdot L$, where u satisfies the standard concavity properties and Inada conditions.[5] In its intertemporal choice, the household faces the discount factor $\rho > 0$. To focus on population and pollution in the absence of production problems, I adopt the simplest formulation for the rest of the model in line with Foster (1973) who assumes that consumption C takes place directly at the cost of environment, i.e., $C = E$. The representative household then chooses emissions $E(t)$ to maximize

$$U = \int_0^\infty u\big[E(t)/L(t)\big]L(t)e^{-\rho t}dt = \int_0^\infty u\big[E(t)/L(t)\big]e^{-\int_0^t\{\rho - n[S(\tau)]\}d\tau}dt, \quad (4)$$

subject to (3). The mechanism of the model is the following: by choosing the optimal path for $E(t)$, the household determines $S(t)$, which in turn dictates the optimal population growth rate $n(t)$ and the optimal population $L(t)$. Finally, per capita emissions $E(t)/L(t)$ are determined.

Because the discount factor in (4) is not constant, I apply the virtual time technique suggested by Uzawa (1968). Let us denote

$$\Delta(t) \equiv \int_0^t \{\rho - n\big[S(\tau)\big]\}d\tau$$

to get $\frac{d\Delta(t)}{dt} = \rho - n[S(t)]$ and $dt = \frac{d\Delta(t)}{\rho - n[S(t)]}$. The problem can now be rewritten in virtual time as:

$$U = \int_0^\infty \frac{u(E/L)}{\rho - n(S)} \cdot e^{-\Delta} \cdot d\Delta,$$

$$\mathring{S} \equiv \frac{dS}{d\Delta} = \frac{dS}{dt}\frac{dt}{d\Delta} = \frac{E - \delta(S)}{\rho - n(S)},$$

where $E \equiv E[\Delta(t)]$, $S \equiv S[\Delta(t)]$, $L \equiv L[\Delta(t)]$. This concave problem with constant discount factor can be solved in virtual time by using standard methods (Benveniste and Scheinkman 1982). Given that both the population size L and its growth rate n depend on the pollution stock S through (1) and (2), the current value Hamiltonian and the necessary conditions become:

$$H(S, E, \lambda) = \frac{1}{\rho - n(S)}\{u(E/L) + \lambda(\Delta)\big[E - \delta(S)\big]\},$$

$$\frac{\partial H(S, E, \lambda)}{\partial E} = 0 \iff -u'(E/L) = \lambda(\Delta) \cdot L, \quad (5)$$

$$\mathring{\lambda} \equiv \frac{d\lambda(\Delta)}{d\Delta} = -\frac{\partial H}{\partial S} + \lambda(\Delta), \quad (6)$$

$$\lim_{\Delta \to \infty} \lambda(\Delta)e^{-\Delta}S = 0. \quad (7)$$

---

[5] Krutilla (1967) and Barbier (2003).

Taking the derivative in (6) and rearranging one gets

$$\mathring{\lambda}/\lambda = -(1/\rho - n)\{n'H/\lambda - (\delta' + \rho - n)\}. \tag{8}$$

To eliminate λ, one can follow the usual procedure by taking the derivative of (5) in terms of (virtual) time. These derivatives are denoted by $\mathring{E} \equiv dE/d\Delta$ and $\mathring{L} \equiv dL/dT$. To simplify the analysis, let us adopt the *CIES* utility function $u(E/L) = [(E/L)^{1-\theta}]/(1-\theta)$, $\theta \neq 1$ with $u'' \cdot (E/L)/u' = -\theta$ to give

$$\mathring{\lambda}/\lambda = -\theta \mathring{E}/E + (\theta - 1)\mathring{L}/L, \tag{9}$$

which together with (8) gives $\mathring{E}/E = [1/\theta(\rho - n)]\{-n'H/\lambda - (\delta' + \rho - \theta n)\}$, where $\mathring{L}/L = n/(\rho - n)$ is applied. Substituting the expression $-n'H/\lambda = [n'/(\rho - n)][\theta E/(\theta - 1) - \delta]$ and noting $\mathring{E} = \dot{E}/(\rho - n)$ one finally derives

$$\frac{\dot{E}}{E} = \frac{1}{\theta}\left\{\frac{n'}{\rho - n}\left[\frac{\theta E}{\theta - 1} - \delta\right] - (\delta' + \rho - \theta n)\right\}. \tag{10}$$

The non-linear equations (3) and (10) supply the solution to the model. The phase lines become:

$$\frac{\dot{E}}{E} = 0 \quad \Leftrightarrow \quad E = \frac{\theta - 1}{\theta}\left\{\delta + \frac{\rho - n}{n'}(\delta' + \rho - \theta n)\right\}, \tag{11a}$$

$$\dot{S} = 0 \quad \Leftrightarrow \quad E = \delta. \tag{11b}$$

In the (S, E)-space, the shape of the $\dot{S} = 0$-line is that of δ, i.e., inverted U with $\delta(0) = \delta(\tilde{S}) = 0$ (Fig. 2). The shape of the $\dot{E} = 0$-line depends on the value of θ. Because Hall has argued that empirical elasticities tend to be large (Hall 1988), I assume $\theta > 1$, but nothing essential is changed if $\theta < 1$ is assumed instead. Even for $\theta > 1$, there is variety in the shape of the $\dot{E} = 0$-line. The following is the sufficient condition for the existence of at least one interior steady state:

Lemma 1 *If $\delta'(0) + \theta n(0) > \rho$ and $\delta'(\tilde{S}) + \theta n(\tilde{S}) < \rho$ then the problem has at least one steady state $S^* \subset (0, \tilde{S})$.*

Proof In the (S, E)-space the $\dot{S} = 0$-line hits the S-axis at $S = 0$ and at $S = \tilde{S}$. For $S = 0$ and $S = \tilde{S}$, (11a) then becomes $\dot{E} = 0 \Leftrightarrow E = \frac{\theta-1}{\theta}\{\frac{\rho-n}{n'}(\delta' + \rho - \theta n)\}$. By assumption, $\theta - 1 > 0$, $\rho - n > 0$ and $n' < 0$. Graphically, if $\delta'(0) + \theta n(0) > \rho$ and $\delta'(\tilde{S}) + \theta n(\tilde{S}) < \rho$, the $\dot{E} = 0$-line lies below the $\dot{S} = 0$-line for $S = 0$ and above it for $S = \tilde{S}$ (Fig. 2). By continuity, the $\dot{E} = 0$-line intersects the $\dot{S} = 0$-line at least once. □

To comprehend, consider marginal emissions. If consumed tomorrow, emissions are discounted by ρ. If consumed today, it adds to the pollution stock S and produces a change in abatement $\delta'(S)$ and population $n(S)$. If the sum of the latter two is larger, consumption today pays. The first unit of emission is consumed if $\delta'(0) +$

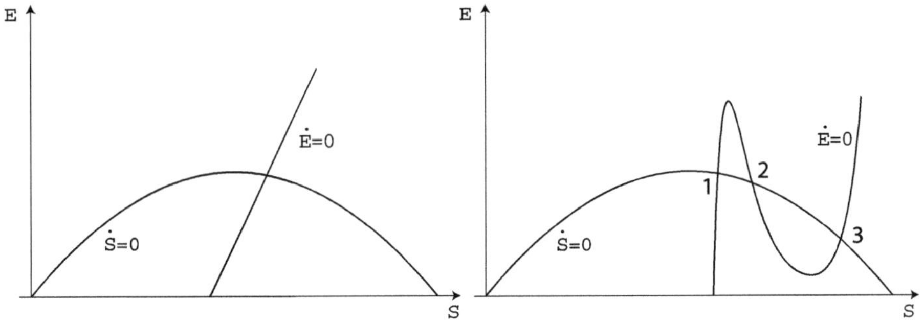

Fig. 2 The phase diagrams of the model

$\theta n(0) > \rho$. On the other hand, if $\delta'(\tilde{S}) + \theta n(\tilde{S}) < \rho$ it never pays to pollute until the carrying capacity $\tilde{S}$.

Depending upon the properties of the population growth function (1), the $\dot{E} = 0$-line may be non-linear and the model may have several steady states; I assume that the number of the steady states is either one or three as shown in Fig. 2. The local stability analysis in Appendix shows that the single steady state is a saddle with stable manifolds running from the North-West and South-East, as the left panel of Fig. 2 illustrates. If the number of the steady states is three (the right panel of Fig. 2), then the first and third are saddles but the second is an unstable focus or node. The following lemma characterizes all saddle-stable steady states:

Lemma 2 *Inefficient under-accumulation of the pollutant is not possible.*

Proof Equations (11a) and (11b) imply that in a steady state

$$\frac{\theta - 1}{\theta}\left\{\delta + \frac{\rho - n}{n'}\left[\delta' + \rho - \theta n\right]\right\} = \delta. \tag{12}$$

The transversality condition is $\lim_{\Delta\to\infty}\{\lambda(\Delta)e^{-\Delta}S(\Delta)\} = 0$. Because the model tends to the steady state, S and $n(S)$ go to constants S^* and $n(S^*)$. In a steady state, $\mathring{E} = 0$ so that (9) implies $\mathring{\lambda}/\lambda = (\theta - 1)\mathring{L}/L$, which is a constant in the steady state. The transversality condition then requires $(\theta - 1)\mathring{L}/L - 1 < 0$. Because $\mathring{L}/L = n/(\rho - n)$, we get $(\theta - 1)n(S^*)/(\rho - n(S^*)) - 1 < 0$ and further

$$\rho - \theta n\left(S^*\right) > 0. \tag{13}$$

Arranging and using (12) we get $\rho - \theta n = \frac{n'}{(\rho-n)(\theta-1)}\delta - \delta' > 0$. Because $\frac{n'}{(\rho-n)(\theta-1)}\delta < 0$, it must be $\delta'(S^*) < 0$. Therefore, the steady state is located on the downwards sloping part of the $\dot{S} = 0$-line. □

4 Sustainability and Technical Progress

The Brundtland Commission 1987 defines sustainable development as a development that "meets the needs of the present without compromising the ability of future generations to meet their own needs" (WCED 1987). This definition refers to non-decreasing consumption or non-decreasing utility, concepts also used by most economists (for a review, see Pezzey 1992). With the positive check present, the concept of sustainability needs a redefinition:

Definition An optimal path is sustainable if it provides non-decreasing consumption for a non-decreasing population.

Thus, an optimal path can lose sustainability either because per capita consumption decreases or because population decreases.

Consider first a steady state. Recall that $E = C$. The growth rate of the per capita consumption is $\gamma_{C/L} = \dot{E}/E - \dot{L}/L$. In the steady state, E is constant so that $\gamma_{C/L} = -\dot{L}/L = -n(S^*)$. Three alternatives are possible. For $n(S^*) > 0$, the population keeps increasing and per capita consumption decreasing. For $n(S^*) = 0$, both the population and per capita consumption are constants. For $n(S^*) < 0$, an ever-decreasing population enjoys ever-increasing per capita consumption. Note that this steady state implies $\lim_{t\to\infty} L(t) = 0$ so that, asymptotically, the size of the population vanishes to zero. Thus, of the above alternatives, only $n(S^*) = 0$ is sustainable.

Which of the above cases realizes? First note that the a priori assumptions $\rho > 0$ and $\rho - n(S) > 0$ pose no explicit limit to $\operatorname{sign} n(S^*)$. Another candidate that would limit $\operatorname{sign} n(S^*)$ is the transversality condition in (13) but for the suitable values of ρ and θ it can hold for positive and negative values of $n(S^*)$. Thus, in the steady state S^* the optimal population may be constant, increasing, or decreasing because the utilitarian objective functional $\int_0^\infty u(E/L)Le^{-\rho t}dt$ may take its maximum both at high E/L and low L or vice versa. Therefore, it may well be optimal to increase consumption at the cost of population.

Some optimists argue, however, that technical progress ultimately warrants sustainability (Neumayer 1999, for example). To see whether this optimism is supported by the model, let $A(t)$ be the available technology at time t and assume that technical progress is exogenously running at rate x so that $A(t) = e^{xt}$ for $A(0) = 1$. Further, let technical progress be consumption augmenting in the meaning that, at every instant of time t, we have $C = e^{xt}E$ implying that for given emissions it is possible to consume more than before (Krautkraemer 1985). Per capita consumption then becomes

$$C/L = e^{xt}E/L. \tag{14}$$

Per capita consumption C/L grows at rate $\gamma_{C/L} = \dot{E}/E + x - n$. In a steady state, $\dot{E}/E = 0$, so that $\gamma_{C/L}$ is positive if $x > n(S^*)$. It is thus *possible* to have growing per capita consumption and growing population together. However, positive popu-

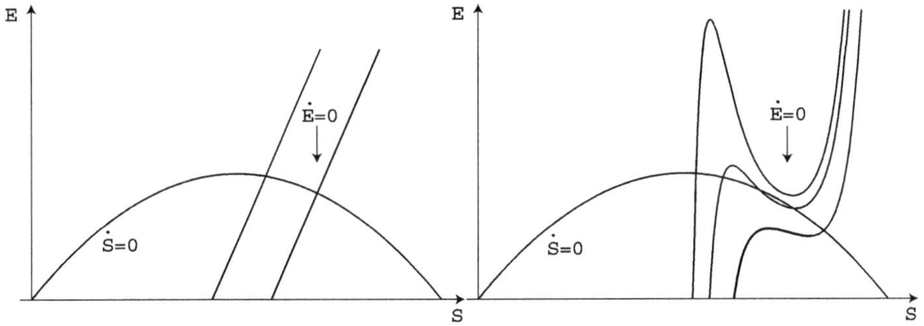

Fig. 3 Technical progress shifts the $\dot{E} = 0$-*line* down and increases the steady state pollution S^*

lation growth is by no means warranted. To see why, apply (14) to (4)–(9) to derive

$$\frac{\dot{E}}{E} = 0 \quad \Leftrightarrow \quad E = \frac{\theta - 1}{\theta}\left\{\delta + \frac{\rho - n}{n'}\left[\delta' + (\theta - 1)x + (\rho - \theta n)\right]\right\}. \qquad (15)$$

The derivative of (15) in terms of x is:

$$\left.\frac{\partial E}{\partial x}\right|_{\dot{E}=0} = \frac{(\theta - 1)^2(\rho - n)}{\theta n'} < 0.$$

Therefore, the $\dot{E} = 0$-line shifts down as the pace of technical progress increases (Fig. 3).

To comprehend, note that in the equilibrium, the *negative* utility effect of a marginal emission through an increase in S and a decrease in population growth, and its *positive* utility effect through an increase in consumption are equal and further emissions are rejected. Technical progress increases the positive consumption effect and larger emission are accepted. Given the biologically determined $\hat{S}$, it is then more likely that $\hat{S} < S^*$ and $n(S^*)$ is negative. Therefore, contrary to conventional wisdom, we find that technical progress does not necessarily save us because it makes extra consumption and emission pay.

To stipulate $\gamma_{C/L} = \dot{E}/E + x - n$ during the transitional period, write

$$\frac{\dot{E}}{E} = \frac{n'}{(\rho - n)(\theta - 1)}\left\{\dot{S} + \frac{1}{\theta}\left[\delta - \frac{(\rho - n)(\theta - 1)}{n'}\left[\delta' + \rho - \theta n + (\theta - 1)x\right]\right]\right\},$$

where the leftmost element is the positive difference of the $\dot{S} = 0$ and $E = 0$-lines indicating $\dot{E}/E < 0$ along the north-western saddle path. Further, as earlier, we have $\lim_{S \to S^*} \dot{E}/E = 0$. Therefore, the sign of $\lim_{S \to S^*} \gamma_{C/L}$ depends on the sign of $x - n(S^*)$. In particular, for $n(S^*) < 0$ we have $x - n(S^*) > 0$ for all x and $\lim_{S \to S^*} \gamma_{C/L} = \dot{E}/E + x - n > 0$ implying that per capita consumption increases as the economy approaches the steady state $n(S^*) < 0$.[6] The following proposition summarizes the results:

Proposition *If the optimal population growth in the steady state is negative, then per capita consumption increases as the economy approaches the steady state. A high rate of technical progress increases the probability for negative steady state population growth.*

5 Parametric Examples

Consider the current pollution-population situation. The population on our planet is larger than ever and increasing, and many specialists argue that we are running out of food supply, that air pollution increases, and that global warming is already on its way. The evidence in Sect. 2 indicates that some signals of the positive check are already available. The parametric examples of this Section try to illustrate the this situation and to give some ideas how our demographic and environmental future looks like.

The abstract style of the model naturally makes its parametric presentation difficult but not impossible.[7] Let us start with the assumption that the carrying capacity of the environment $\tilde{S}$ takes some arbitrary value, say $\tilde{S} = 1000$. Since this value refers to a complete disappearance of life, it seems that, in spite of some alarming signals, this situation is not very close yet. Thus, let the current pollution stock be $S(0) = 250$ which is one quarter of $\tilde{S} = 1000$. Further, let $n(S(0)) = 0.005$, indicating that the current (initial) population growth rate is 0.5%. Next, assume that the environmental mortality is high enough to push the population growth below zero if pollution reaches three quarters of $\tilde{S} = 1000$, implying that the critical value is $\hat{S} = 750$.

Other parameters of the model are adapted such that they are in line with the benchmark values above. Consider the population function given in (1) and Fig. 1. The parametric examples provided here concentrate on cases A and B which refer to linear and threshold population function respectively (Fig. 1). These functions are specified as

$$n(S) = \beta - \eta S, \tag{16a}$$

$$n(S) = \beta - \frac{\alpha}{1 + (\mu S)^{-\gamma}}, \tag{16b}$$

[6]The slope of the entire time path for $\gamma_{C/L}$ depends on $\lim_{S \to 0} \gamma_{C/L}$. This and the cases $n(S^*) > 0$ and $n(S^*) = 0$ are not considered for shortness.

[7]The critical obstacle, preventing a full calibration on real data is that, on order to focus on population and pollution, no production function is specified in the model. The main simplification is that the stock of capital (another state variable) is left away, which makes the optimization procedure much simpler and the phase portrait much more intuitive.

where the latter is one of the simplest expressions to produce a threshold function. In the linear case (16a), the demographic response to pollution is given by a single parameter, $\eta > 0$, whereas this response is more complicated in the threshold case as $\beta - \alpha$ gives the lowest population growth reached, $\mu > 0$ multiplies the effect of pollution such that a large value of μ leads to negative population growth at low concentrations and $\gamma > 0$ gives the curvature of the threshold function with high values referring to a highly curved shape and severity of the mortality crisis. In both (16a) and (16b), the parameter β gives the autonomous population growth, i.e., the population growth rate which prevails in a complete the absence of pollution ($S = 0$).

To meet the benchmark values $n(S(0)) = n(250) = 0.005$ and $\tilde{S} = 750$, the parameters of the linear case (16b) must be $\beta = 0.0075$ and $\eta = 0.00001$. In the non-linear case (16b), the autonomous population growth rate $\beta \approx 0.005$ directly warrants $n(S(0)) = n(250) = 0.005$. The choice $\alpha \approx 0.020$ indicates that the lowest population growth reached is -0.15%, a value that seems reasonable even though it can not be derived from the benchmark values above. Further, if $\gamma = 8$ then $\mu \approx 0.00116$ warrants the property $\tilde{S} = 750$ for the threshold case. The top panel in Fig. 4 illustrates.

The abatement function $\delta(S)$ takes the standard logistic formula

$$\delta(S) = rS\left(1 - \frac{S}{\tilde{S}}\right), \tag{17}$$

in which r is the intrinsic rate of annual decay with an assumed value of $r = 0.175$. Conventional values $\theta = 4$ and $\rho = 0.03$ describe the preferences (see Barro and Sala-i-Martin 1995, for example). Two rates of technical progress are assumed, namely $x = 0.00$ and $x = 0.02$. All parameters are collected to Table 1. To summarize, there will be four cases, the linear case without and with technical progress, referred to as A00 and A02, and the threshold case without and with technical progress, referred to as B00 and B02 respectively.[8]

Table 2 reports the main steady-state results of the parameterized model and Fig. 4 illustrates, showing that all cases have a single steady state S^*. Table 2 shows that, in the absence of technical progress we have $S^* < \hat{S}$ both in linear and threshold cases while, in the presence of technical progress, the opposite is true, i.e., $S^* > \hat{S}$. Thus, technical progress makes the steady state population to decrease both in the linear and threshold case, the half-life times being 3623 and 181 years, respectively (Table 2). On the other hand, the steady state population increases by 0.17% or 0.45% if there is no technical progress, doubling in 409 and 153 years (Table 2).

The depicted off-steady-state paths for population in Fig. 5 (left) show that population first rises from the initial value $L(0) = 1$ in all cases, continues to rise for A00 and B00, almost levels-off for A02 and starts to decrease for B02. In

[8]Calculations are performed by Mathematica 7.0. Time-elimination method is used to derive the saddle paths (Mulligan and Sala-i-Martin 1991).

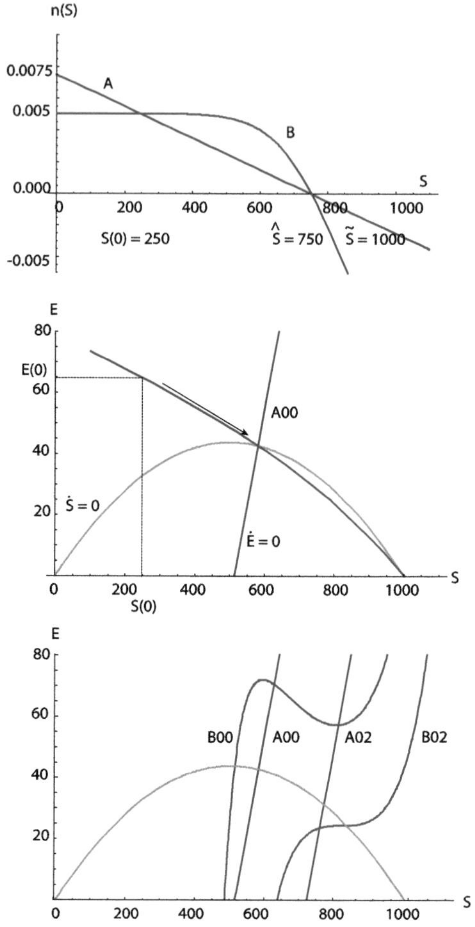

Fig. 4 The parametric population growth functions *A* and *B* (*top*), the phase diagram for the case *A*00 (*middle*), and a combined phase diagram for all cases (*bottom*)

the latter case, the initial population $L = 1$ is reached after 160 years. The difference between the population projections is prominent, indeed. The per capita emission (per capita consumption) paths, instead, are rather similar initially. But after some hundred years, per capita emissions along B02 start to rise as the number of people decreases, meeting the proposition in Sect. 4. Thus, in B02, it is optimal to choose higher and higher per capita consumption at the cost of lower and lower number of people. Note also that per capita consumption almost levels-off in A02. Given that population levels-off as well, A02 almost meets the sustainability as defined in Sect. 4, but only by change.[9] To summarize, the parametric example provided here imply that the utility-maximizing path with positive check may take a large variety of consumption-population combinations depending upon

[9]Note, that in the presence of technical progress, however, the case with rising per capita consumption and rising population in the steady state is possible, see Sect. 4.

Table 1 The parameters of the model

Parameter	Linear A	Threshold B
$\tilde{S}$	1000	1000
r	0.175	0.175
β	0.0075	0.005
η	0.00001	
α		0.020
μ		0.00116
γ		8
θ	4	4
ρ	0.03	0.03
x	0.00 or 0.02	0.00 or 0.02

the parameters of the population growth functions and on the rate of technical progress.

Several extensions of the current model are both necessary and possible. Maybe the first of them would be to include a realistic production function in order to see how the role of population (labor) as a factor of production changes the results. A more realistic version of technical progress would take this progress as a response to environmental degradation and overpopulation. This paper assumes that all technical progress is consumption-augmenting, but technical progress may also

Table 2 The results of the model

Results	A00	A02	B00	B02
x	0.00	0.02	0.00	0.02
S^*	580.7	769.1	545.8	833.8
$n(S^*)$	0.17%	−0.019%	0.45%	−0.38%
Doubling/half-life (years)	409	3623	153	181

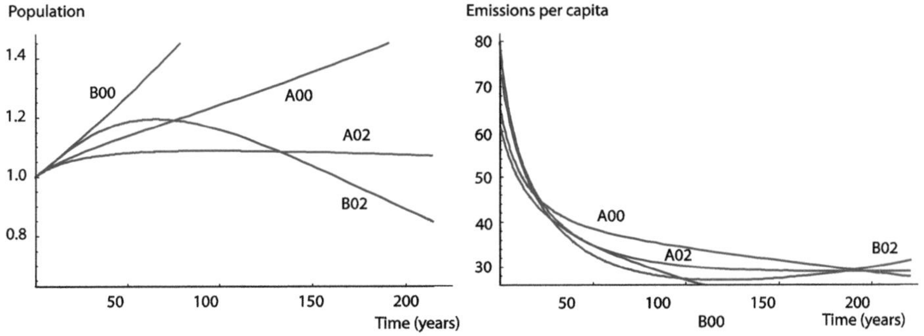

Fig. 5 The parametric time paths for population and per capita consumption

save lives and increase longevity. It may be "dirty" or "clean" as it may either increase the number of polluting product variants or rise the consumer's utility value of constant-pollution goods (Palokangas 2012; Smulders et al. 2011). Furthermore, since new ideas and *R&D* are positively related to the population size (Kremer 1993), the economies should be better equipped to tackle the environmental problems in the future. Public policies to support technical progress should be modeled as well. On the consumer's side, the option to choose between dirty and clean goods and between lower and higher birth rates should enrich the model. All these extensions should be made in the future. Nevertheless, it seems that they will not change the basic implication of the model, namely that there is a fundamental trade-off between per capita consumption and population in the long run optimization. This trade-off arises because per capita consumption and population are both valuable to man and because emissions increase the former but decrease the latter. The trade-off implies that sustainable paths may appear, but are not warranted, indicating that optimality and sustainability may conflict for the reasonable parameters and functional specifications of the model.

6 Discussion

Any article on sustainable growth is, more or less, a wake-up call. Broadly speaking, one wants to predict what happens if the currently shown disturbing behavior continues and if environmental concerns are not taken seriously. Currently, some people suffer and die for environmental reasons but the vast majority consumes ever more. If, however, pollution-related mortality remains tolerable, the worrisome conclusion is that, in the real world as well as in the model, the incentives for a change in economic behavior may not be sufficient.

The long run consumer optimization with endogenous pollution and endogenous population implies that utility maximization may take place at the cost of human lives. Solow has suggested that "The theory of optimal growth… is thoroughly utilitarian in conception. It is also utilitarian in the narrow sense that social welfare is (usually) defined as the sum of the utilities of different individuals or generations" (Solow 1974). In the case of endogenous pollution and endogenous population, this utilitarianism may take an extreme expression: a path that ultimately leads to self-imposed extinction may still be optimal. Naturally, a different result would have been derived if positive population were posed as an a priori constraint on optimization. However, an emerging empirical evidence suggests that there already is an increase in mortality because of environmental reasons. Therefore, as a description of the current situation, the utilitarian approach may not be so distorted after all.

Appendix: Local Stability of the Steady States

Lets write $\dot{S} = \varphi(S, E)$ and $\dot{E} = \phi(S, E)$. In a steady state it hods $\dot{S} = \dot{E} = 0$ implying

$$\delta + \delta' S + \rho - \theta n = \frac{n'}{(\rho - n)(\theta - 1)} \delta S. \tag{18}$$

The Jacobian of the model is

$$J = \begin{bmatrix} \varphi_S & \varphi_E \\ \phi_S & \phi_E \end{bmatrix}.$$

As evaluated around a steady state, its elements become

$$\begin{aligned}
\varphi_S &= -\left(\delta + \delta' S\right), \\
\varphi_E &= 1, \\
\phi_S &= \frac{E}{\theta} \left\{ \begin{matrix} \frac{-n''(\rho-n)-(n')^2}{(\rho-n)^2} [\frac{\theta E}{1-\theta} + \delta S] \\ -\frac{n'}{\rho-n} [\delta + \delta' S] - [2\delta' + \delta'' S - \theta n'] \end{matrix} \right\}, \\
\phi_E &= \frac{1}{\theta} \left\{ \frac{n'}{\rho - n} \left[\frac{\theta E}{\theta - 1} + \delta S \right] - \left[\delta + \rho + \delta' S - \theta n\right] \right\} + \frac{E}{\theta} \left\{ \frac{n'}{\rho - n} \frac{\theta}{\theta - 1} \right\} \\
&= \frac{n' E}{(\rho - n)(\theta - 1)},
\end{aligned}$$

in which the last row is derived by using (12) and (11b). Because ϕ_E contains the undefined second derivative $n(S)$, we write

$$\begin{aligned}
\mathrm{DET}\, J &= \varphi_S \cdot \phi_E - \phi_S \cdot \varphi_E \\
&= \left[\left(-\frac{\varphi_S}{\varphi_E} \right) - \left(-\frac{\phi_S}{\phi_E} \right) \right] (-\varphi_E) \cdot \phi_E.
\end{aligned}$$

The expression in the square brackets is the difference in the slopes of the phase lines $\dot{S} = 0$ and $\dot{E} = 0$ and $(-\varphi_E) \cdot \phi_E = -\frac{n'E}{(\rho-n)(\theta-1)}$ is positive for all $E > 0$. In steady states 1 and 3 the slope of the $\dot{E} = 0$-line is steeper than that of the $\dot{S} = 0$-line (see Fig. 2) making the square brackets negative. Thus, $\mathrm{DET}\, J < 0$ and these steady states are saddles. In steady state 2 the slope of the $\dot{E} = 0$-line is smaller (possibly negative) than the slope of the $\dot{S} = 0$-line and the value of the square brackets is positive. The trace of the Jacobian is

$$\begin{aligned}
\mathrm{TR}\, J &= \varphi_S + \phi_E \\
&= -\left(\delta + \delta' S\right) + \frac{n' E}{(\rho - n)(\theta - 1)} \\
&= -\left(\delta + \delta' S\right) + \frac{n' \delta S}{(\rho - n)(\theta - 1)}
\end{aligned}$$

$$= -\left(\delta + \delta' S\right) + \frac{n'}{(\rho - n)(\theta - 1)} \cdot \frac{(\rho - n)(\theta - 1)}{n'} \cdot \left(\delta + \delta' S + \rho - \theta n\right)$$
$$= \rho - \theta n > 0.$$

Therefore, this steady state is an unstable node or focus.

References

Barbier, E. B. (2003). The role of natural resources in economic development. *Australian Economic Papers*, *42*(2), 253–272.

Barro, R. J., & Sala-i-Martin, X. (1995). *Economic growth* (1st edn.). New York: McGraw-Hill.

Benveniste, L., & Scheinkman, J. (1982). Duality theory of dynamic optimization models of economics: the continuous time case. *Journal of Economic Theory*, *27*, 1–19.

Brunekreef, B., & Holgate, S. T. (2002). Air pollution and health. *Lancet*, *360*(9341), 1233–1242.

Chay, K. Y., & Greenstone, M. (2003). The impact of air pollution on infant mortality: evidence from geographic variation in pollution shocks induced by a recession. *Quarterly Journal of Economics*, *118*, 1121–1167.

Currie, J., & Neidell, M. (2004). Air pollution and infant health: what can we learn from California's recent experience? *Quarterly Journal of Economics*, *120*, 1003–1030.

Dasgupta, P. S., & Heal, G. (1974). The optimal depletion of exhaustible resources. *Review of Economic Studies*, *41*, 3–28.

Deschênes, O., & Greenstone, M. (2011). Climate change, mortality, and adaptation: evidence from annual fluctuations in weather in the U.S. *American Economic Journal: Applied Economics*, *3*(4), 152–185.

Evans, M. F., & Smith, K. V. (2005). Do new health conditions support mortality—air pollution effects? *Journal of Environmental Economics and Management*, *50*, 496–518.

Foster, B. A. (1973). Optimal consumption planning in a polluted environment. *Economic Record*, *49*(128), 534–545.

Gosling, S. N., Lowe, J. A., McGregor, G. R., Pelling, M., & Malamud, B. D. (2009). Associations between elevated atmospheric temperature and human mortality: a critical review of the literature. *Climatic Change*, *92*, 299–341.

Hall, R. E. (1988). Intertemporal substitution in consumption. *Journal of Political Economy*, *96*(2), 339–357.

IPPC (2003). *Climate change, impacts, adaption, and vulnerability. Intergovernmental panel on climate change, the third assessment*. Cambridge: Cambridge University Press.

Keeler, E., Spence, M., & Zeckhauser, R. (1971). The optimal control of pollution. *Journal of Economic Theory*, *4*, 19–34.

Krautkraemer, J. A. (1985). Optimal growth, resource amenities and the preservation of natural environments. *Review of Economic Studies*, *52*, 153–170.

Kremer, M. (1993). Population growth and technological change: one million B.C. to 1990. *The Quarterly Journal of Economics*, *108*(3), 681–716.

Krutilla, J. V. (1967). Conservation reconsidered. *American Economic Review*, *57*(4), 777–786.

Logan, W. P. D. (1953). Mortality in London fog incident, 1952. *Lancet*, *261*(6755), 336–338.

Lutz, W., O'Neill, F., & MacKellar, L. (2005). *Population and climate change*. New York: Cambridge University Press.

Malthus, T. R. (1914). *An essay on the principle of population*. London: Dent. 1973 reprint.

Meadows, D., Meadows, D., Zahn, E., & Milling, P. (1972). *The limits to growth*. London: Universe Books.

Mulligan, C. B., & Sala-i-Martin, X. (1991). *A note on the time-elimination method for solving recursive dynamic economic models* (Working Paper 116). National Bureau for Economic Research.

Nemery, B., Hoet, P. H. M., & Nemmar, A. (2001). The Meuse Valley fog 1930: an air pollution disaster. *Lancet*, *357*, 704–708.

Neumayer, E. (1999). *Weak versus strong sustainability*. Cheltenham Glos: Edward Elgar.

Palokangas, T. (2012). *Clean versus dirty economic growth* (Discussion Paper 648:2012). Department of Economics, University of Helsinki.

Pezzey, J. C. (1992). *Sustainable development concepts* (Working Paper 11425). World Bank.

Pezzey, J. C., & Withagen, C. A. (1998). The rise, fall, and sustainability of capital-resource economies. *Scandinavian Journal of Economics*, *100*(2), 513–527.

Pitcher, H., Ebi, K., & Brenkert, A. (2008). Population health model for integrated assessment models. *Climatic Change*, *88*(1), 35–57.

Plourde, C. (1972). A model of waste accumulation disposal. *Canadian Journal of Economics*, *5*(1), 119–125.

Pope, C. A., & Dockery, D. W. (2006). Health effects of fine particulate air pollution: lines that connect. *Journal of Air Waste Management Association*, *56*(6), 709–742.

Pope, C. A., Burnett, R. T., Thun, M. J., Calle, E. E., Krewski, D., Ito, K., & Thurston, G. D. (2002). Lung cancer, cardiopulmonary mortality, and long-term exposure to fine particulate air pollution. *Journal of the American Medical Association*, *287*(9), 1132–1141.

Raaschou-Nielsen, O., Andersen, Z. J., Hvidberg, M., Jensen, S. S., Ketzel, M., Sorensen, M., & Tjonneland, A. (2011). Lung cancer incidence and long-term exposure to air pollution from traffic. *Environmental Health Perspective*, *119*(6), 860–865.

Samet, J. M., Dominici, F., Curriero, F. C., Coursac, I., & Zeger, S. L. (2000). Fine particulate air pollution and mortality in 20 U.S. cities, 1987–1994. *New England Journal of Medicine*, *343*, 1742–1749.

Scheers, H., Mwalili, S. M., Faes, C., Fierens, F., Nemery, B., & Nawrot, T. S. (2011). Does air pollution trigger infant mortality in western Europe? A case-crossover study. *Environmental Health Perspectives*, *119*, 1017–1022.

Smulders, S., & Gradus, R. (1996). Pollution abatement and long-term growth. *European Journal of Political Economy*, *12*, 505–532.

Smulders, S., Bretschger, L., & Egli, H. (2011). Economic growth and the diffusion of clean technologies: explaining environmental Kuznets curves. *Environmental and Resource Economics*, *49*(1), 79–99.

Solow, R. (1974). Intergenerational equity and exhaustive resources. *Review of Economic Studies*, *41*, 29–45.

Stiglitz, J. (1974). Growth with exhaustible natural resources: efficient and optimal growth paths. *Review of Economic Studies*, *41*, 123–137.

Tahvonen, O., & Salo, S. (1996). Nonconvexities in optimal pollution accumulation. *Journal of Environmental Economics and Management*, *31*, 160–177.

Tanser, F. C., Sharp, B., & Le Sueur, D. (2003). Potential effect of climate change on malaria transmission in Africa. *Lancet*, *362*, 1792–1798.

Uzawa, H. (1968). Time preference, the consumption function, and optimum asset holdings. In J. Wolfe (Ed.), *Value, capital, and growth* (Chap. 21). Chicago: Aldine.

WCED (1987). *Our common future*. New York: Oxford University Press.

WHO (2003). *Climate change and human health, risks and responses*, Geneva: World Health Organization.

WHO (2004). *Meta-analysis of time-series studies and panel studies of particulate matter (PM) and ozone (O3)*. Copenhagen: World Health Organization, Regional Office for Europe.

WHO (2011). *Air quality and health* (Fact Sheet Number 313). Updated September 2011. http://www.who.int/mediacentre/factsheets/fs313/en/index.html.

WHO (2012). Health and environment linkages initiative. http://www.who.int/heli/risks/climate/climatechange/en/.

Optimal Proportions in Growth Trends of Resource Productivity

Alexander Tarasyev and Bing Zhu

1 Introduction

The paper is devoted to the problem of optimizing trends in resources productivity and balancing investment in economy's dematerialization with sustainable growth of the consumption index. The problem is considered within the classical approach (Solow 1970; Shell 1969) of construction of economic growth models. The main new element in the proposed model is a price formation mechanism which reflects possibility of rapid growth of prices on exhausting resources. Growing prices negatively influence on the consumption index which should be maximized in the model as the basic element of the utility function. Let us note that the stated problem has in its background very important concerns of the modern society with respect to the current world resource utilization. The recent statistics (IPCC 2007; OECD 2011) shows rapid increase of natural resource consumption, especially, in the following components: fossil energy (oil, natural gas, oil), ferrous metals (iron ore, etc.), non-ferrous metals (bauxite, etc.), non-metalliferous minerals (lime), biomass (wood, etc.). Taking into account the limitations of natural resources, at least, of its assured

A. Tarasyev (✉)
IMM UrB RAS, S. Kovalevskaya str. 16, Ekaterinburg, Russia
e-mail: tam@imm.uran.ru

A. Tarasyev · B. Zhu
IIASA, Laxenburg, Austria

A. Tarasyev
e-mail: tarasiev@iiasa.ac.at

B. Zhu
e-mail: zhu@iiasa.ac.at

B. Zhu
Tsinghua University, Beijing, China
e-mail: bingzhu@tsinghua.edu.cn

J. Crespo Cuaresma et al. (eds.), *Green Growth and Sustainable Development*, Dynamic Modeling and Econometrics in Economics and Finance 14,
DOI 10.1007/978-3-642-34354-4_3,

part, the problem of raising resource efficiency and even reducing resource consumption becomes extremely significant. Nowadays, a comprehensive research is being implemented on material flow analysis (MFA) by international (EUROSTAT, IPCC, OECD, World Resources Institute) and national (Germany, the Netherlands, the United States, Japan, China) research and policy making organizations. Material flow analysis is a systematic assessment of the flows and stocks of materials within a system defined in space and time. It connects the sources, the pathways, and the intermediate and the final sinks of a material. The method is an attractive decision-support tool in resource management, waste management, and environmental management.

In this paper, we supplement this research and develop the model of dynamic optimization of investment process in improving resource productivity within the economic growth theory (Arrow 1985; Ayres and Warr 2009; Barro and Sala-i Martin 1995; Crespo-Cuaresma et al. 2010; Gordon et al. 1988; Grossman and Helpman 1991). Particularly, the construction of the model inherits elements of economic growth models introduced in Ane et al. (2007a, 2007b), Ayres et al. (2009), Krasovskii and Tarasyev (2009), Krasovskii et al. (2008), Sanderson et al. (2010), Tarasyev and Watanabe (2001), Tarasyev et al. (2002), Watanabe et al. (2009) and Kryazhimskii and Watanabe (2004). Let us mention here papers (Aseev et al. 2010; Feichtinger et al. 2006; Hutschenreiter 1995) which are devoted to different aspects of economic growth modeling and conceptually are close to our approach. The model dynamics includes production, current material use and cumulative material consumption as main phase variables. Growing trend in production is given exogenously by the exponential term generated by such production factors as capital and labor. Material use is introduced as a production factor in the production function of the Cobb–Douglas type. The main control variable is presented by investment in raising resource productivity in the current period.

It is assumed that prices on materials due to exhaustion are growing rapidly to infinity when the cumulative material consumption is close to the available (assured) stock. In the balance equation both growth and decline trends are taken into account: the growth trend in the consumption index is stimulated by the production growth and the decline trend is caused by raising costs of materials and expenditures directed on improvement of resource productivity.

The problem is to find the optimal proportion of investment in the dynamic process with maximization of the utility function given as the integrated consumption index over trajectories of the economic system. The model is examined within the framework of the Pontryagin maximum principle (Pontryagin et al. 1962) with special characteristics of infinite horizon (Aseev and Kryazhimskiy 2007). Specific features of the corresponding Hamiltonian system are examined within the qualitative theory of differential equations (Hartman 1964). In our analysis we use constructions of dynamic programming and the theory of generalized solutions of Hamilton–Jacobi equations (Bellman 1957; Krasovskii and Krasovskii 1995; Subbotin 1995; Rockafellar 2004). The range of model parameters is indicated for existence and uniqueness of a steady state. The steady state plays the role of the optimal steady solution and its proportions can be used as an economic standard for the first approximation of solution of the optimal control problem. It is shown that at the steady state the optimal level of investment in resource productivity provides reduction in resource consumption and raise of its efficiency, and establish a reasonable balance between investment and consumption. The main output of the implemented analysis and modeling is construction of investment strategies in economy's dematerialization and improvement of resource productivity.

2 Model Description

We assume that the model dynamics is evolved in time t on the infinite horizon $[0, +\infty)$. The main phase variables of the model are presented by the current production $y = y(t)$, the resource use $m = m(t)$ and the cumulative resource consumption which is introduced as the integrated material use

$$M = M(t) = \int_0^t m(s)ds. \tag{1}$$

Initial values for the resource use and the cumulative resource consumption are defined by the levels $m(0) = m^*$ and $M(0) = M^*$, respectively.

Resource productivity in period t is denoted by the symbol $z = z(t)$ and is given by the following expression

$$z(t) = \frac{y(t)}{m(t)}. \tag{2}$$

Further, for convenience we use in several relations the value of resource intensity—the inverse value to productivity

$$Z(t) = \frac{1}{z(t)} = \frac{m(t)}{y(t)}. \tag{3}$$

Price Formation Mechanism In the definition of the price formation mechanism the basis is provided by the concept of raising prices $p(t)$ on natural resources in the case of their limitation or exhaustion. It is assumed that prices are growing according to the inversely proportional rule of resource exhaustion

$$p = p(t) = p_0\left(1 - \frac{M(t)}{M_0}\right)^{-\gamma}, \quad \gamma \geq 0. \tag{4}$$

Here the parameter γ is the elasticity coefficient of the price formation mechanism, the symbol M_0 stands for the limitation of natural resources, and the symbol p_0 denotes the initial price on natural resources. Formula (4) envisages that price $p(t)$ can grow rapidly to infinity according to the hyperbolic law when the integrated material use $M(t)$ reaches its limitation M_0.

Balance Equation In the balance equation it is taken into account that production $y(t)$ in period t is shared between consumption $c(t)$, from the one hand, and the growing cost of natural resources $p(t)m(t)$ plus investment $s(t)$ in improving the resource productivity, from the other hand,

$$y(t) = c(t) + p(t)m(t) + s(t). \tag{5}$$

Let us assume that there exists an upper bound s^0 for investment $s(t)$, i.e. $0 \leq s(t) \leq s^0 < y(t)$. Deducing the consumption intensity $c(t)/y(t)$ from (5) through the resource intensity $m(t)/y(t)$ we obtain the following relation

$$\frac{c(t)}{y(t)} = 1 - p(t)\frac{m(t)}{y(t)} - u(t). \tag{6}$$

Here the symbol $u(t)$ stands for the investment intensity

$$u(t) = \frac{s(t)}{y(t)}. \tag{7}$$

We assume that there exists an upper bound u^0 for the investment intensity $u(t)$, so $u(t) \leq u^0$.

Production Function The exponential production function of the Cobb-Douglas type is chosen for the first version of the model

$$y(t) = ae^{bt}m^{\alpha}(t), \quad a > 0, b \geq 0, 0 \leq \alpha < 1. \tag{8}$$

Here the parameter a is a scale factor; the growth rate b indicates the growth process of production $y(t)$ due to development of basic production factors such as capital, labor, technology, etc.; the symbol α denotes the elasticity coefficient of natural resources. We assume the diminishing return to scale of natural resources as a production factor, $0 \leq \alpha < 1$. Principally, one can assume in formula (8) that production $y(t)$ is normalized with respect to the labor growth and stands for per capita production.

Consumption Intensity Let us obtain the formula for the consumption intensity expressed through the resources consumption $m(t)$, $M(t)$, by substituting relations of the price formation mechanism (4) and the production function (8) to relation (6)

$$\frac{c(t)}{y(t)} = 1 - \frac{1}{a}p_0e^{-bt}\left(1 - \frac{M(t)}{M_0}\right)^{-\gamma} m^{1-\alpha}(t) - u(t). \tag{9}$$

Model Dynamics Let assume that the relative raise in the resource productivity $z(t)$ is proportional to the portion of the assigned investment $u(t)$ which can be interpreted as investment in "green" technology (see Grossman and Helpman 1991)

$$\frac{1}{z(t)}\frac{dz(t)}{dt} = \beta u(t), \quad \beta \geq 0. \tag{10}$$

Here the parameter β describes the effectiveness of investment $u(t)$ in raising the resource productivity. Taking into account the definition (2) of the resource productivity $z(t)$ one can obtain the following presentation for its rate

$$\frac{1}{z(t)}\frac{dz(t)}{dt} = \frac{dy(t)}{y(t)} - \frac{dm(t)}{m(t)}. \tag{11}$$

The last equation means that the rate of the resource productivity can be decomposed into two components: the production rate and the rate of the resource consumption. Developing this formula further on the basis of the presentation for the production function (8) we get the following relation

$$\frac{1}{z(t)}\frac{dz(t)}{dt} = b - (1-\alpha)\frac{dm(t)}{m(t)}. \tag{12}$$

Finally, combining formulas (10) and (12) we derive the equation for the rate of the resource consumption

$$\frac{dm(t)}{m(t)} = \frac{1}{1-\alpha}\big(b - \beta u(t)\big). \tag{13}$$

Equation (13) shows that the rate of the resource consumption is influenced by the production growth rate b and can be reduced only by investment $u(t)$ in raising the resource productivity. Let us note that if investment is equal to zero, $u(t) = 0$, then the rate of the resource consumption should be proportional to the production growth rate b.

To develop the model dynamics further, let us introduce the following change of variables

$$x_1(t) = e^{-b\gamma t/(1-\alpha-\gamma)}\left(1 - \frac{M(t)}{M_0}\right)^{\gamma}, \qquad x_2(t) = e^{-bt/(1-\alpha-\gamma)}m(t). \tag{14}$$

We derive the differential equations for the model dynamics by differentiating variables $x_1(t)$, $x_2(t)$ in time t and taking into account (1)–(2), (8), (13). We obtain the following differential equations which form the basic model dynamics

$$\frac{dx_1(t)}{dt} = -\gamma\left(\frac{b}{1-\alpha-\gamma}\frac{1}{M_0}x_1^{-1/\gamma}(t)x_2(t)\right)x_1(t), \tag{15}$$

$$\frac{dx_2(t)}{dt} = -\frac{1}{1-\alpha}\left(\frac{b\gamma}{1-\alpha-\gamma} + \beta u(t)\right)x_2(t), \tag{16}$$

$$x_1(0) = 1, \qquad x_2(0) = m^*. \tag{17}$$

Initial conditions (17) mean that the variable $x_1(t)$ is an analogue of the cumulative resource consumption $M(t)$, and the variable $x_2(t)$ is equivalent to the current resource use $m(t)$.

It is important to remind that the control variable $u(t)$ in the model dynamics (15)–(17) is subject to constraints

$$0 \leq u(t) \leq u^0 < 1. \tag{18}$$

Logarithmic Consumption Index Using variables $x_1(t)$, $x_2(t)$ we introduce the logarithmic consumption index in period t

$$\begin{aligned}
\ln c(t) &= \ln y(t) + \ln\left(1 - \frac{p_0}{\alpha}\frac{x_2^{1-\alpha}(t)}{x_1(t)} - u(t)\right) \\
&= \ln\left(ae^{bt}m^{\alpha}(t)\right) + \ln\left(1 - \frac{p_0}{\alpha}\frac{x_2^{1-\alpha}(t)}{x_1(t)} - u(t)\right) \\
&= \ln\left(ae^{(1-\gamma)bt/(1-\alpha-\gamma)}\right) + \alpha\ln x_2(t) + \ln\left(1 - \frac{p_0}{\alpha}\frac{x_2^{1-\alpha}(t)}{x_1(t)} - u(t)\right).
\end{aligned}$$

Considering the model dynamics (15)–(17) on the time horizon $[0, T)$, $T \leq +\infty$, we introduce the integrated logarithmic index discounted with the discount rate ρ, $\rho > 0$,

$$\begin{aligned}
J\left(x_1(\cdot), x_2(\cdot), u(\cdot)\right) = \int_0^T e^{-\rho t}\Bigg(\ln a + \frac{(1-\gamma)bt}{1-\alpha-\gamma} \\
+ \alpha\ln x_2(t) + \ln\left(1 - \frac{p_0}{a}\frac{x_2^{1-\alpha}(t)}{x_1(t)} - u(t)\right)\Bigg)dt,
\end{aligned} \tag{19}$$

as the utility function for the optimal control problem.

3 A Model Interpretation

Let us suppose that there are two agents: the (representative) resource owner; and the central planner that represents all the other agents in the economy. The consumption of the resource owners is not included in the social welfare function. There are two possible justifications as follows:

1. The resource owners are foreigners that repatriate their profits abroad. The local central planner does not mind the foreigners' welfare.
2. The resource owners are domestic, they consume their profits, but their proportion in the population (among the voters) is so small that the central planner ignores their welfare.

In both cases, the central planner bases the social welfare function only on the consumption of the rest of the population.

The inverse supply function (4) can be derived from the resource owners' behavior e.g. as follows:

1. Assume that the (representative) resource owner faces extraction cost f, which is an increasing and convex function of the use of resources, M:

$$f = f(M), \qquad f'(M) > 0, \qquad f''(M) > 0.$$

The resource owner maximizes its profit $pM - f(M)$, where M is the quantity, p is the price and $f(M)$ is the extraction cost of natural resources. This leads to the profit maximization condition

$$p = f'(M).$$

2. Let's put the cost function in the parametric form

$$f(M) = \theta\left(1 - \frac{M}{\varphi}\right)^{1-\gamma},$$

where θ, φ and γ are parameters. Let us choose the unit of resources so that $\varphi = M_0$, where M_0 is the limit value of the parameter M. This implies

$$p = f'(M) = \frac{(\gamma - 1)\theta}{\varphi}\left(1 - \frac{M}{\varphi}\right)^{-\gamma} = \frac{(\gamma - 1)\theta}{M_0}\left(1 - \frac{M}{M_0}\right)^{-\gamma}.$$

Now the initial price is given by

$$p_0 = p|_{M=M_0} = \frac{(\gamma - 1)\theta}{M_0}.$$

Solving for θ implies

$$\theta = \frac{p_0 M_0}{\gamma - 1}.$$

Plugging this and $\varphi = M_0$ into the cost function implies

$$f(M) = \theta\left(1 - \frac{M}{\varphi}\right)^{1-\gamma} = \frac{p_0 M_0}{\gamma - 1}\left(1 - \frac{M}{M_0}\right)^{1-\gamma} > 0,$$

where $\gamma > 1$. This leads to the supply function

$$p = f'(M) = p_0\left(1 - \frac{M}{M_0}\right)^{-\gamma}.$$

4 Optimal Control Problem

We pose the optimal control problem related to the goal of raising the resource productivity. Namely, the problem is to maximize the utility function (19) over control processes $(x_1(t), x_2(t), u(t))$ of the dynamic system (15)–(17) satisfying the initial conditions $(x_1^0, x_2^0) = (1, m^*)$ and subject to constraints (18) for the control parameter $u(t)$.

Special Case $\gamma = 1$ Let us consider the special case when the elasticity coefficient in the price formation mechanism has the unit value, $\gamma = 1$. In this case the phase variables have the following form

$$x_1(t) = e^{bt/\alpha}\left(1 - \frac{M(t)}{M_0}\right), \qquad x_2(t) = e^{bt/\alpha} m(t). \tag{20}$$

The model dynamics is described by the system of differential equations

$$\begin{aligned} \frac{dx_1(t)}{dt} &= \frac{b}{\alpha} x_1(t) - \frac{1}{M_0} x_2(t), \quad x_1(0) = 1, \\ \frac{dx_2(t)}{dt} &= \left(\frac{b}{\alpha} - \beta u(t)\right) \frac{x_2(t)}{1-\alpha}, \quad x_2(0) = m^* \end{aligned} \tag{21}$$

with the same constraints (18) on the control parameter $u(t)$. The utility function has the form similar to the structure (19)

$$\begin{aligned} J\big(x_1(\cdot), x_2(\cdot), u(\cdot)\big) = \int_0^T & e^{-\rho t}\Bigg(\alpha \ln x_2(t) \\ &+ \ln\left(1 - \frac{p_0}{a} \frac{x_2^{1-\alpha}(t)}{x_1(t)} - u(t)\right)\Bigg) dt. \end{aligned} \tag{22}$$

5 The Hamiltonian of the Optimal Control Problem

Let us introduce the Hamiltonian function for the optimal control problem (21)–(22)

$$\begin{aligned} \widetilde{H}(x, u, t, \widetilde{\psi}) = e^{-\rho t}\Bigg(& \alpha \ln x_2 + \ln\left(1 - \frac{p_0}{a} \frac{x_2^{1-\alpha}}{x_1} - u\right) \\ &+ \left(\frac{b}{\alpha} x_1 - \frac{1}{M_0} x_2\right) \widetilde{\psi}_1 + \left(\frac{b}{\alpha} - \beta u\right) \frac{x_2 \widetilde{\psi}_2}{1-\alpha}\Bigg). \end{aligned}$$

Here symbol x denotes the vector of phase variables $x = (x_1, x_2)$, parameter $\widetilde{\psi} = (\widetilde{\psi}_1, \widetilde{\psi}_2)$ is the vector of adjoint variables for the phase variables x_1, x_2. Implementing the following change of variables

$$\psi_1(t) = e^{\rho t} \widetilde{\psi}_1(t), \qquad \psi_2(t) = e^{\rho t} \widetilde{\psi}_2(t)$$

we obtain the expression for the stationary Hamiltonian $H(x,u,\psi)$, where $H(x, u,\psi)=e^{-\rho t}\widetilde{H}(x,u,t,\widetilde{\psi})$

$$H(x,u,\psi)=\alpha\ln x_2+\ln\left(1-\frac{p_0}{a}\frac{x_2^{1-\alpha}}{x_1}-u\right)$$
$$+\left(\frac{b}{\alpha}x_1-\frac{1}{M_0}x_2\right)\psi_1+\left(\frac{b}{\alpha}-\beta u\right)\frac{x_2\psi_2}{1-\alpha}, \tag{23}$$

where $\psi=(\psi_1,\psi_2)$.

6 The Maximized Hamiltonian

Let us maximize the stationary Hamiltonian (23) with respect to the control parameter u. Using methods of convex analysis one can show that the Hamiltonian H is strictly concave with respect to this parameter. Therefore (see Krasovskii and Tarasyev 2009), three maximum regimes for the control parameter u, and, respectively, for the maximized Hamiltonian may take place.

The first regime corresponds to the zero value of the control parameter, $u=0$. For this regime the maximized Hamiltonian has the following form

$$H_1(x,\psi)=\alpha\ln x_2+\ln\left(1-\frac{p_0}{a}\frac{x_2^{1-\alpha}}{x_1}\right)$$
$$+\left(\frac{b}{\alpha}x_1-\frac{x_2}{M_0}\right)\psi_1+\frac{b}{\alpha(1-\alpha)}x_2\psi_2, \tag{24}$$

where symbols x and ψ denote vectors (x_1,x_2) and (ψ_1,ψ_2), respectively.

The second regime arises at the upper bound for the control parameter, $u=u^0$. The maximized Hamiltonian in this case is presented by the following relation

$$H_2(x,\psi)=\alpha\ln x_2+\ln\left(1-\frac{p_0}{a}\frac{x_2^{1-\alpha}}{x_1}-u^0\right)$$
$$+\left(\frac{b}{\alpha}x_1-\frac{x_2}{M_0}\right)\psi_1+\left(\frac{b}{\alpha}-\beta u^0\right)\frac{x_2\psi_2}{1-\alpha}. \tag{25}$$

The third regime is connected with an intermediate maximum value of the optimal parameter and is determined by the maximum condition

$$\frac{\partial H(x,u,\psi)}{\partial u}=\frac{-1}{1-(p_0/a)(x_2^{1-\alpha}/x_1)-u}-\frac{\beta x_2\psi_2}{1-\alpha}=0. \tag{26}$$

Resolving the maximum condition (26) with respect to the control parameter u we obtain the relation for the intermediate maximum value

$$u^*=1-\frac{p_0}{a}\frac{x_2^{1-\alpha}}{x_1}+\frac{1-a}{\beta x_2\psi_2}. \tag{27}$$

It is clear from formula (26) that the adjoint variable ψ_2 is negative, $\psi_2 < 0$, for the intermediate regime. The maximized Hamiltonian for the intermediate regime has the following form

$$H_3(x,\psi) = \alpha \ln x_2 - 1 + \ln\left(-\frac{1-\alpha}{\beta x_2 \psi_2}\right) + \frac{b}{\alpha} x_1 \psi_1 - \frac{x_2}{M_0}\psi_1 + \left(\frac{b}{\alpha} - \beta + \frac{\beta p_0}{a}\frac{x_2^{1-\alpha}}{x_1}\right)\frac{x_2\psi_2}{1-\alpha}. \tag{28}$$

7 The Hamiltonian Systems

Let us compile the Hamiltonian systems for the obtained three control regimes, $i = 1, 2, 3$, basing on the general constructions of the Pontryagin maximum principle

$$\dot{x}_j(t) = \frac{\partial H_i(x(t), \psi(t))}{\partial \psi_j}, \qquad \dot{\psi}_j(t) = \rho \psi_j(t) - \frac{\partial H_i(x(t), \psi(t))}{\partial x_j},$$

where $j = 1, 2$.

For providing economic interpretations of the Hamiltonian dynamics we introduce the following change variables $z_1 = \psi_1 x_1$, $z_2 = \psi_2 x_2$ for costs of material consumption x_1, x_2, by prices ψ_1, ψ_2, respectively. For the Hamiltonian dynamics of costs z_1, z_2, in the case ($i = 3$) of the intermediate optimal control $u = u^*$ one can obtain the following system of differential equations

$$\begin{aligned}
\dot{x}_1(t) &= \frac{b}{\alpha} x_1(t) - \frac{1}{M_0} x_2(t),\\
\dot{x}_2(t) &= \frac{x_2(t)}{1-\alpha}\left(\frac{b}{\alpha} - \frac{1-\alpha}{z_2(t)} - \beta\left(1 - \frac{p_0}{a}\frac{x_2^{1-\alpha}(t)}{x_1(t)}\right)\right),\\
\dot{z}_1(t) &= \left(\rho - \frac{x_2(t)}{M_0 x_1(t)}\right) z_1(t) + \frac{p_0 \beta x_2^{1-\alpha}(t)}{a(1-\alpha)x_1(t)} z_2(t),\\
\dot{z}_2(t) &= \frac{x_2(t)}{M_0 x_1(t)} z_1(t) + \left(\rho - \frac{p_0 \beta x_2^{1-\alpha}(t)}{a x_1(t)}\right) z_2(t) - \alpha.
\end{aligned} \tag{29}$$

Let us denote the right-hand parts of the Hamiltonian dynamics (29) by symbols $G_j(x,z)$, $j = \overline{1,4}$.

8 Existence of Steady State

We are interested in existence of steady states for the Hamiltonian dynamics (29) which are connected with the structure of the optimal solution of the posed optimal control problem. The equilibrium conditions for steady states are presented by the system of algebraic equations

$$G_j(x,z)=0, \quad j=\overline{1,4}. \tag{30}$$

Let us note that the steady state solution can be considered as the "ideal" equilibrium state of the economic growth model at which the variables of material consumption x_1, x_2, and their costs z_1, z_2 keep constant equilibrium values.

Proposition 1 *The solution for the steady state equations* (30) *exists and can be found analytically under the regularity conditions*

$$\beta > \rho > \frac{b}{\alpha}. \tag{31}$$

Remark 1 The first inequality in (31) means that the effectiveness coefficient β of investment in raising the resource productivity should be greater than the discount rate ρ. The second inequality in (31) presumes that the discount rate ρ is larger than the growth rate b of production factors since elasticity coefficient α is less than one, $\alpha < 1$.

Proof of Proposition 1 Under conditions (31) the steady state $P^* = (x_1^*, x_2^*, z_1^*, z_2^*)$ as the solution of equilibrium equations (30) has the following analytical form

$$\begin{aligned}
x_1^* &= \frac{\alpha}{bM_0}\left(\frac{\beta\rho b p_0 M_0}{a(\alpha\rho - b)((\rho - b) + \alpha(\beta - \rho))}\right)^{1/\alpha},\\
x_2^* &= \left(\frac{\beta\rho b p_0 M_0}{a(\alpha\rho - b)((\rho - b) + \alpha(\beta - \rho))}\right)^{1/\alpha},\\
z_1^* &= \frac{\alpha(\rho(1-\alpha) + \alpha\beta - b)}{\alpha\rho(1-\alpha)(\beta - \rho) + b(\alpha\beta - b)},\\
z_2^* &= -\frac{\alpha\rho(1-\alpha)}{\alpha\rho(1-\alpha)(\beta - \rho) + b(\alpha\beta - b)}.
\end{aligned} \tag{32}$$

Let us note that due to the regularity conditions (31) all coordinates of solution (32) have the property of well-posedness $x_1^* > 0$, $x_2^* > 0$, $z_1^* > 0$, $z_2^* < 0$. □

Proposition 2 *The value of the optimal control u^* at the steady state P^** (32) *is estimated by the relation*

$$u^* = \frac{b}{\alpha\beta}. \tag{33}$$

Proof Substituting formulas (32) for coordinates of the steady state to relation (27) of optimal control in the intermediate regime one can get the necessary expression (33) of optimal control at the steady state. □

Remark 2 Due to regularity conditions (31) the value of the optimal control u^* is located in the proper range $0 < u^* < 1$. It means that it is reasonable to make an assumption that the upper bound u^0 for the control parameter u should satisfy to the following condition

$$\frac{b}{\alpha\beta} \leq u^0 < 1. \tag{34}$$

Remark 3 Under regularity conditions (31) a nontrivial (nonzero) steady state exists only for dynamics (29) corresponding to the maximized Hamiltonian for the intermediate regime (28). There are no nontrivial steady states for dynamics corresponding to the maximized Hamiltonian functions for the zero regime (24) and for the upper bound regime (25). Thus, there exists the unique nontrivial steady state for the series of the maximized Hamiltonian functions (24)–(28).

9 Qualitative Analysis of Model Solutions at the Steady State

Let us analyze properties of the main model solutions at the steady state P^* and prove that they have realistic trends of sustainable development of model trajectories.

Proposition 3 *At the steady state the current resource use $m(t)$ is declining to zero according to the exponential law*

$$m(t) = x_2^* e^{bt/\alpha}. \tag{35}$$

The cumulative resource consumption $M(t)$ increases up to the limit level M_0 of natural resources according to the logistic growth law

$$M(t) = M_0\big(1 - x_1^* e^{-bt/\alpha}\big). \tag{36}$$

Proof Formulas (35)–(36) follow immediately from dynamic equations (21) integrated with the optimal control (33). □

Proposition 4 *The price parameter $p(t)$ generated by the price formation mechanism* (4) *increases exponentially at the steady state*

$$p(t) = \frac{p_0}{x_1^*} e^{bt/\alpha}. \tag{37}$$

Proof Exponential growth property of price $p(t)$ (37) is generated by the logistic growth law (36) of the cumulative resource consumption $M(t)$. □

Finally, we determine optimal levels of production y and consumption index c.

Proposition 5 *Production y^* and consumption index c^* have strictly positive values at the steady state*

$$y^* = a(x_2^*)^\alpha = \frac{\beta\rho b p_0 M_0}{(\alpha\rho - b)((\rho - b) + \alpha(\beta - \rho))} > 0, \tag{38}$$

$$\begin{aligned} c^* &= a(x_2^*)^\alpha \left(1 - \frac{b}{\alpha\beta}\right) - \frac{b}{\alpha} p_0 M_0 \\ &= \frac{b^2 p_0 M_0((\alpha\beta - b) + (1-\alpha)(\beta - \rho))}{\alpha(\alpha\rho - b)(\rho - b + \alpha(\beta - \rho))} > 0. \end{aligned} \tag{39}$$

$$\nu^* = \frac{c^*}{y^*} = \frac{b((\alpha\beta - b) + (1-\alpha)(\beta - \rho))}{\alpha\beta\rho}, \tag{40}$$

where symbol ν^* denotes the consumption intensity evaluated at the steady state P^*.

Proof Relations (38)–(40) follow from the balance equation (5), relation (8) for the production function, and formulas (32) for the steady state under the regularity condition (31). □

Let us calculate the absolute value of investment s^* and derive proportions between consumption c^* and investment s^*.

Proposition 6 *The optimal absolute value of investment s^* is given by the formula*

$$s^* = u^* y^* = \frac{b^* \rho p_0 M_0}{\alpha(\alpha\rho - b)((\rho - b) + \alpha(\beta - \rho))}. \tag{41}$$

Proportion between consumption c^* and investment s^* is given by the ratio

$$w^* = \frac{c^*}{s^*} = \frac{(\rho - b) + \alpha(\beta - \rho)}{\rho}. \tag{42}$$

10 Sensitivity Analysis of Steady State

Basing on analytical solutions (33), (38)–(39) for the steady state coordinates one can implement sensitivity analysis of the equilibrium solution with respect to model parameters.

Proposition 7 *The optimal level of control u^** (33) (*investment intensity*) *has the following trends*:

(1) *it increases when the growth parameter b in the production function increases*;
(2) *it decreases when the elasticity coefficient α of the production function increases*;

Fig. 1 Investment $s^* = s^*(\alpha, \beta)$

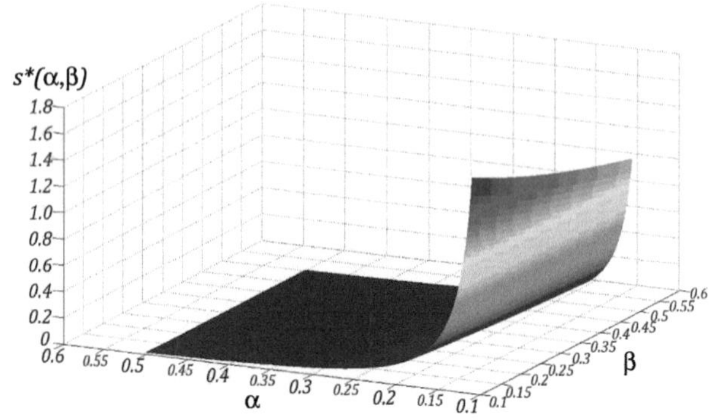

(3) *it decreases when the effectiveness coefficient β of investment in the resource productivity increases*;
(4) *it does not depend on the discount rate ρ.*

Proof These properties are direct conclusions from proportionality conditions for the optimal control (33). □

Let us estimate trends of the absolute value of investment.

Proposition 8 *The optimal absolute value of investment s^** (41) *has the following trends*:

(1) *it increases when the growth parameter b in the production function increases*;
(2) *it decreases when the elasticity coefficient α of the production function increases*;
(3) *it decreases when the effectiveness coefficient β of investment in the resource productivity increases*;
(4) *it decreases when the discount rate ρ increases.*

Proof The first three trends are quite evident and follow from the growing and declining properties of numerator and denominator in (41). To show the fourth trend, we calculate the partial derivative with respect to the discount parameter ρ and verify that it is negative

$$\frac{\partial s^*}{\partial \rho} = -b^2 p_0 M_0 \frac{(b(\rho - b + \alpha(\beta - \rho)) + \rho(1-\alpha)(\alpha\rho - b))}{\alpha(\alpha\rho - b)^2(\rho - b + \alpha(\beta - \rho))^2} < 0.$$ □

The optimal absolute value of investment s^* as function in parameters α and β is depicted on Fig. 1.

Proposition 9 *The optimal level of production y^** (38) *has the following trends*:

(1) *it increases when the growth parameter b in the production function increases*;

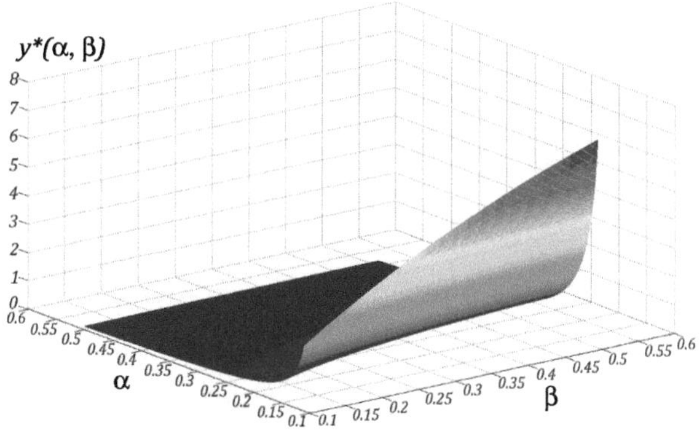

Fig. 2 Investment $y^* = y^*(\alpha, \beta)$

(2) *it decreases when the elasticity coefficient* α *of the production function increases*;

(3) *in the case when the elasticity coefficient* $\alpha \leq 0.5$, *production* y^* *increases when the effectiveness coefficient* β *of investment in the resource productivity increases*;
in the case when the elasticity coefficient $\alpha > 0.5$, *both growing and declining trends are feasible*;

(4) *it decreases when the discount rate* ρ *increases.*

Proof The indicated trends are obtained in analysis of positive and negative signs of the partial derivatives of production y^* (38) with respect to the corresponding model parameters. The estimation of trends (1), (2) and (3) is similar to calculations in the proof of the previous statement. To demonstrate the third property we calculate the partial derivative of production y^* with respect to the effectiveness parameter β

$$\frac{\partial y^*}{\partial \beta} = \frac{\rho b p_0 M_0((1-\alpha)\rho - b)}{(\alpha\rho - b)(\rho - b + \alpha(\beta - \rho))^2}.$$

Due to the regularity condition (31) the obtained derivative is strictly positive in the case when $\alpha \leq 0.5$. In the opposite case, both signs positive and negative are feasible. □

The optimal level of production y^* as function in parameters α and β is depicted on Fig. 2. Let us consider behavior of proportion between consumption c^* and investment s^* at the steady state P^*.

Proposition 10 *Proportion* w^* *between consumption* c^* *and investment* s^* *has the following trends at the steady state*:

(1) *it decreases when the growth parameter* b *in the production function increases*;

(2) *it increases when the elasticity coefficient* α *of the production function increases*;

(3) *it increases when the effectiveness coefficient* β *of investment in the resource productivity increases*;

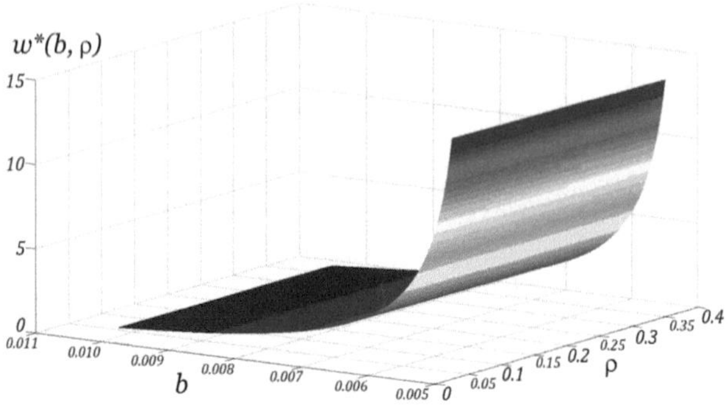

Fig. 3 Proportion w^* between consumption c^* and investment s^* as function of parameters b and ρ, $w^* = w^*(b, \rho)$

(4) *it decreases when the discount rate ρ increases.*

Proof One can easily calculate partial derivatives of proportion c^*/s^* and estimate their signs

$$\frac{\partial w^*}{\partial b} = -\frac{1}{\rho} < 0, \qquad \frac{\partial w^*}{\partial \alpha} = 1 > 0,$$

$$\frac{\partial w^*}{\partial \beta} = \frac{1}{\rho} > 0, \qquad \frac{\partial w^*}{\partial \rho} = -\frac{\beta - b}{\rho^2} < 0.$$

The obtained signs in the corresponding derivatives prove the statement. □

The optimal level of the proportion w^* between consumption c^* and investment s^* as function in parameters b and ρ is depicted on Fig. 3.

The implemented sensitivity analysis shows that the model synthetic solutions at the steady state demonstrate quite adequate trends and provide reasonable proportions of equilibrium sustainable development. This observation creates a basis for application of the model to the real data analysis and forecasting procedures.

11 Conclusion

The paper is devoted to construction of the model for optimizing investment in raising resource productivity and establishing a proper balance for production and consumption. One of the basic features of the model is implementation of the price formation mechanism which generates rapidly growing prices for exhausting resources. The balance is formed in the consumption index which negatively depends on growing prices on materials. The optimal control problem for the investment process is posed and solved within the Pontryagin maximum principle. Properties of the corresponding Hamiltonian systems are analyzed in order to obtain stationary optimal model solutions. Sufficient conditions for specific range of the model parameters are indicated providing the existence result for the steady state of the

Hamiltonian system which is interpreted as the stationary optimal solution. Analytical formulas are derived for coordinates of the steady state and its derivatives with respect to model parameters. The related sensitivity analysis is given for the stationary equilibrium solution.

Acknowledgements The research is supported by the Russian Fund of Basic Research (Grants 11-01-0042-a, 11-01-12088-ofi-m-2011, 11-01-12112-ofi-m-2011), by the Program for the Sponsorship of Leading Scientific Schools (Grant NSCH-64508.2010.1), by the Program of the Presidium of RAS "Dynamic Systems and Control Theory", by the Program of the Presidium of RAS No. 38П (Project 12-П-1-1038), by the Project of the Ural Branch of RAS "Socio-Economic Development of Regions: Forecasting and Optimal Control" (Grant 12-П-7-1001), and the International Institute for Applied System Analysis (IIASA) within the NSFC-IIASA Collaboration Project "Optimization of Resource Productivity for Sustainable Economic Development".

References

Ane, B. K., Tarasyev, A. M., & Watanabe, C. (2007a). Construction of nonlinear stabilizer for trajectories of economic growth. *Journal of Optimization Theory and Applications*, *134*(2), 303–320.

Ane, B. K., Tarasyev, A. M., & Watanabe, C. (2007b). Impact of technology assimilation on investment policy: dynamic optimization and econometric identification. *Journal of Optimization Theory and Applications*, *134*(2), 321–338.

Arrow, K. J. (1985). *Collected papers: Vol. 5. Production and capital.* Cambridge: Harvard University Press.

Aseev, S. M., & Kryazhimskiy, A. V. (2007). *Proceedings of the Steklov Institute of Mathematics: Vol. 257. The Pontryagin maximum principle and optimal economic growth problems.* Moscow: Pleiades Publishing.

Aseev, S., Besov, K., & Kaniovski, S. (2010). *Optimal endogenous growth with exhaustible resources* (Interim Report IR-10-011). IIASA.

Ayres, R. U., & Warr, B. (2009). *The economic growth engine: how energy and work drive material prosperity*. Cheltenham Glos: Edward Elgar.

Ayres, R., Krasovskii, A. A., & Tarasyev, A. M. (2009). Nonlinear stabilizers of economic growth under exhausting energy resources. In *Proceedings of the IFAC CAO'09 workshop on control applications of optimization*, University of Jyuvaskyla.

Barro, R. J., & Sala-i-Martin, X. (1995). *Economic growth.* New York: McGraw-Hill.

Bellman, R. (1957). *Dynamic programming*. Princeton: Princeton University Press.

Crespo-Cuaresma, J., Palokangas, T., & Tarasyev, A. (Eds.) (2010). *Dynamic modeling and econometrics in economics and finance. Dynamic systems, economic growth and the environment.* Berlin: Springer.

Feichtinger, G., Hartl, R. F., Kort, P. M., & Veliov, V. M. (2006). Capital accumulation under technological progress and learning: a vintage capital approach. *European Journal of Operation Research*, *172*(1), 293–310.

Gordon, R. B., Koopmans, T. C., Nordhaus, W., & Skinner, B. J. (1988). *Toward a new iron age? A study of patterns of resource exhaustion.* Cambridge: Harvard University Press.

Grossman, G. M., & Helpman, E. (1991). *Innovation and growth in the global economy.* Cambridge: MIT Press.

Hartman, Ph. (1964). *Ordinary differential equations*. New York: Wiley.

Hutschenreiter, G., Kaniovski, Y., & Kryazhimskii, A. (1995). *Endogenous growth, absorptive capacities and international R&D spillovers* (Working Paper 95-92). IIASA.

IPCC (2007). *Fourth assessment report: climate change 2007 (AR4).* Cambridge: Cambridge University Press. www.ipcc.ch/publications_and_data/publications_and_data_reports.shtml.

Krasovskii, A. A., & Tarasyev, A. M. (2009). High-precision algorithms for constructing optimal trajectories via solving Hamiltonian systems. In *Proceedings of IFAC CAO'09 workshop on control applications of optimization*, University of Jyuvaskyla.

Krasovskii, A. N., & Krasovskii, N. N. (1995). *Control under lack of information*. Boston: Birkhauser.

Krasovskii, A., Kryazhimskiy, A., & Tarasyev, A. (2008) Optimal control design in models of economic growth. In P. Neittaanmki, J. Priaux & T. Tuovinen (Eds.), *Evolutionary methods for design, optimization and control* (pp. 70–75). Barcelona: CIMNE.

Kryazhimskii, A. V., & Watanabe, C. (Eds.) (2004). *Optimization of technological growth*. Kanagawa: Gendaitosho.

OECD (2011). *Policy instruments for sustainable materials management*. Paris: OECD Publishing. http://www.oecd.org/env/resourceproductivityandwaste/49804504.pdf.

Pontryagin, L. S., Boltyanskii, V. G., Gamkrelidze, R. V., & Mishchenko, E. F. (1962). *The mathematical theory of optimal processes*. New York: Interscience.

Rockafellar, R. T. (2004). Hamilton-Jacobi theory and parametric analysis in fully convex problems of optimal control. *Journal of Global Optimization*, *28*, 419–431.

Sanderson, W., Tarasyev, A. M., & Usova, A. A. (2010). Capital vs education: assessment of economic growth from two perspectives. In *Preprints of the IFAC symposium on nonlinear control systems (NOLCOS 2010)*, Bologna.

Shell, K. (1969). Applications of Pontryagin's maximum principle to economics. *Mathematical Systems Theory and Economics*, *1*, 241–292.

Solow, R. M. (1970). *Growth theory: an exposition*. New York: Oxford University Press.

Subbotin, A. I. (1995). *Systems and control: foundations and applications. Generalized solutions of first-order PDE: the dynamical optimization perspective*. Boston: Birkhauser.

Tarasyev, A. M., & Watanabe, C. (2001). Optimal dynamics of innovation in models of economic growth. *Journal of Optimization Theory and Applications*, *108*, 175–203.

Tarasyev, A. M., Watanabe, C., & Zhu, B. (2002). Optimal feedbacks in techno-economic dynamics. *International Journal of Technology Management*, *23*(7/8), 691–717.

Watanabe, C., Shin, J.-H., Heikkinen, J., & Tarasyev, A. (2009). Optimal dynamics of functionality development in open innovation. In *Proceedings of IFAC CAO'09 workshop on control applications of optimization*, University of Jyvaskyla.

Part II
Biodiversity, Abatement and Climate Change

International Biodiversity Management with Technological Change

Tapio Palokangas

1 Introduction

This document considers optimal institutional design of international biodiversity management under lobbying, with a special focus on the following problems. Should biodiversity management be run by an international agency or by individual countries independently? How much authority should this agency get? Are regulatory powers sufficient, or should the international agency have a budget to finance conservation subsidies, for instance?

The framework for this study is based on the following experience. The "international agency" called the European Commission (EC) manages biodiversity and two directives regulate nature conservation in the European Union (EU) (cf. Ostermann 1998):

- Birds Directive 79/409/EEC on the conservation of wild birds;
- Habitats Directive 92/43/EEC on the conservation of natural habitats and of wild fauna and flora.

The Habitats Directive calls for the establishment of a network of designated sites, called Natura 2000, which will consist of sites designated under the Habitats Directive (Special Areas of Conservation, SACS) and the Birds Directive (Special Protection Areas, SPAS). These directives contain annexes with habitats and species listed as being of Community interest, and whose conservation requires the designation of sites by the Member States. A Member State is obliged to guarantee a "Favorable Conservation Status", which is defined in the Habitats Directive, to a Natura 2000 site with the obligations of monitoring and reporting.

Non-governmental organizations (NGOs) play a crucial role in the highly complex political structure of the EU. Weber and Christophersen (2002) describe the

T. Palokangas (✉)
University of Helsinki and HECER, "Economicum" Building, Arkadiankatu 7, P.O. Box 17, 00014 Helsinki, Finland
e-mail: Tapio.Palokangas@helsinki.fi

J. Crespo Cuaresma et al. (eds.), *Green Growth and Sustainable Development*, Dynamic Modeling and Econometrics in Economics and Finance 14,
DOI 10.1007/978-3-642-34354-4_4, © Springer-Verlag Berlin Heidelberg 2013

political influence of the forest-owner associations (CEPF and BNFF) and the environmental NGOs (WWF and Fern) on the process of implementing the EU habitats directive (HD). They highlight the relationship between the involvement of interest groups in the political process and the acceptance of legislation among their members. This document examines the political equilibrium in which the interest groups representing the member countries lobby the Commission over biodiversity management.

There are three reasons why EU policy relies heavily on regulation rather than on other mechanisms to achieve its objectives (Ledoux et al. 2000).

- Until 1987, EU environmental policy lacked a proper legal basis in the founding Treaty of Rome. Consequently, all environmental policies had to rely on the "implied powers" of Article 235 of the Treaty, which stipulated the use of directives and nothing else.
- With the ratification of the 1999 Amsterdam Treaty, the EU can only adopt ecotaxes and other fiscal measures with the unanimous agreement of every state (Jordan 1998). This need for unanimity represents both a huge hurdle to ecological tax reform and a continuing institutional inducement to rely on regulation.
- The founding Member States gave the EU a powerful institutional incentive to regulate wherever possible by vesting it with so few financial resources of its own. From the Commission's perspective, regulation has the benefit of being paid for by private actors in the Member States rather than the EU itself (Majone 1996).

In this document, three cases of biodiversity management are considered:

- There is no such international authority as the Commission.
- The current situation in the EU: regulation by the Commission.
- The Commission gets more authority: it can use subsidies and distribute the costs of these to the member countries.

The comparison of these cases reveals whether or not the Commission's present authority is adequate.

MacArthur and Wilson (1967) show that the total number of species is an increasing function of the habitat area. On the assumption that the number of species yields utility, Swanson (1994), Barbier and Schulz (1997) and Endres and Radke (1999) consider the optimal area of habitat, comparing the benefits of its maintenance with the opportunity cost of using land in production. These authors analyze the effects of an external shock (e.g. a change in trade policy) on biodiversity. Rowthorn and Brown (1999) introduce exogenous technological change into the optimal habitat model, finding that a country with a high discount rate preserves more land when the elasticity of substitution between consumption and species exceeds unity.

The optimal choice of a habitat is merely that of allocating land between conservation and production without abatement investment. This document shows that the introduction of abatement investment leads to the following positive link between biodiversity and technological change. The protection of biodiversity requires transferring land from production to conservation. If this decreases output and the employment of labor, then wages fall. Lower wages encourage abatement investment, speeding up technological change.

To consider the political economy of biodiversity management, lobbying is introduced into the optimal habitat model. This can be examined either by the *all-pay auction model* in which the lobbyist making the greater effort wins with certainty, or the *menu-auction model* in which the lobbyists announce their bids contingent on the politician's actions. In the all-pay auction model, lobbying expenditures are incurred by all the lobbyists before the politician takes an action. A good example of this is Johal and Ulph (2002) in which local interest groups lobby to influence the probability of getting their favorite type of government elected. In the menu-auction model, it is not possible for a lobbyist to spend money and effort on lobbying without getting what he lobbied for. This model is chosen here, because it characterizes better the case in which (i) the international agency's decision variables lobbied over (e.g., regulatory constraints, subsidies) are continuous and (ii) the interest groups obtain marginal improvements in their position by lobbying. In this document, the international agency is self-interested, households love biodiversity, goods are produced from labor and land and biodiversity is an increasing function of habitat land in all countries of the economy.

This paper is organized as follows. Section 2 presents the structure of the economy and Sect. 3 the model for a single country. Section 4 constructs the Pareto optimum for the economy as a reference case. Sections 5 and 6 examine the two alternatives of biodiversity management: direct regulation and conservation subsidies.

2 The Model

Consider an economy with a large number of countries which are placed evenly over the limit [0, 1].[1] All countries produce the same consumption good at the price p. Each country j possesses one unit of labor, of which the amount l_j is devoted to production and the rest z_j to abatement investment, and one unit of land, of which the amount n_j is devoted to production and the rest b_j to conservation:

$$1 = l_j + z_j, \qquad 1 = n_j + b_j. \tag{1}$$

MacArthur and Wilson (1967) show empirically that the number of species expected to survive in an island is proportional to the area of that island. Following Rowthorn and Brown (1999), the area devoted to conservation, b_j, functions like an "island" in the MacArthur-Wilson sense in each country j. Thus, *biodiversity* in the economy, b, can be specified simply as the sum of conserved areas in the economy:

$$b \doteq \int_0^1 b_k dk. \tag{2}$$

[1] If the countries were heterogeneous, then there could be multiple equilibria.

There is a single revenue-maximizing agent (hereafter called country j) that controls all resources in country j. Its utility starting at time T is[2]

$$\int_T^\infty c_j b^\delta e^{-\rho(\theta-T)} d\theta, \quad \delta > 0, \ \rho > 0, \tag{3}$$

where θ is time, ρ the constant rate of time preference, c_j its consumption, b biodiversity, and δ a parameter with the following characterization: the higher δ, the more the households appreciate biodiversity in the economy, b. Because there is no money in the model that would pin down the nominal price level at any time, the monetary unit can be chosen so that the consumer price $(1+\tau)p$, where p is the producer price and τ is the consumption tax, is equal to the externality effect b^δ in the model:

$$(1+\tau)p = b^\delta \quad \text{or} \quad p = b^\delta/(1+\tau). \tag{4}$$

2.1 Technology

When country j develops a new technology, it increases its total factor productivity (TFP) by the constant $a > 1$. Its TFP is then equal to a^{γ_j}, where γ_j is its technology serial number. Given TFP, country j is subject to the CES production function $f(l_j, n_j)$ with constant returns to scale, where l_j (n_j) is the input of labor (land):

$$\begin{aligned} y_j &= a^{\gamma_j} f(l_j, n_j), \qquad f_l > 0, \qquad f_n > 0, \qquad f_{ll} < 0, \\ f_{ln} &= -f_{ll} l_j/n_j, \qquad f_{nn} = -f_{ln} l_j/n_j = f_{ll}(l_j/n_j)^2, \end{aligned} \tag{5}$$

where the subscript l (n) denotes the partial derivative with respect to l_j (n_j). In this one-good economy, total consumption is equal to total production:

$$\int_0^1 c_j dj = \int_0^1 y_k dk. \tag{6}$$

Because the labor (land) market is competitive, the producer real wage (rent) w_j (r_j) is determined by the marginal product of labor (land):

$$w_j = \partial y_j/\partial l_j = a^{\gamma_j} f_l(l_j, n_j), \qquad r_j = \partial y_j/\partial l_j = a^{\gamma_j} f_n(l_j, n_j). \tag{7}$$

Noting (5) and (7), the expenditure shares of land ξ and labor $1-\xi$ are

$$\frac{n_j f_n(l_j, n_j)}{f(l_j, n_j)} \doteq \xi\left(\frac{l_j}{n_j}\right), \qquad \frac{w_j l_j}{y_j} = \frac{l_j f_l(l_j, n_j)}{f(l_j, n_j)} = 1 - \xi\left(\frac{l_j}{n_j}\right). \tag{8}$$

---

[2]With the general form of the utility function, $\int_T^\infty c_j^{1-\beta} b^\delta e^{-\rho(\theta-T)} d\theta$, where $\beta \in [0, 1)$ is a constant, it would be very difficult to find a stationary state in the model.

2.2 Research and Development

The improvement of technology in country j depends on labor devoted to abatement investment in that country, z_j. In a small period of time dt, the probability that abatement investment will lead to development of a new technology with a jump from γ_j to $\gamma_j + 1$ is given by $\lambda z_j dt$, while the probability that abatement investment will remain without success is given by $1 - \lambda z_j dt$, where the constant λ is productivity in abatement investment. Noting (1), this defines a Poisson process χ_j with

$$d\chi_j = \begin{cases} 1 & \text{with probability } \lambda z_j dt, \\ 0 & \text{with probability } 1 - \lambda z_j dt, \end{cases} \qquad z_j = 1 - l_j, \tag{9}$$

where $d\chi_j$ is the increment of the process χ_j. The expected growth rate of productivity a^{γ_j} is given by

$$g_j \doteq E\left[\log a^{\gamma_j+1} - \log a^{\gamma_j}\right] = (\log a)\lambda z_j = (\log a)\lambda(1 - l_j), \tag{10}$$

where E is the expectation operator (cf. p. 59 in Aghion and Howitt 1998).

2.3 The International Agency

The international agency does not observe the level of productivity, a^{γ_j}, but observes the producer real wage w_j and the producer rent r_j in each country j. It is assumed that the only revenue-raising tax is the tax τ on consumption expenditure $p\int_0^1 c_k dk$, where p is the consumption price and c_k consumption in country k.[3] With a subsidy η to abatement investment expenditure $w_j z_j$ and a subsidy s to expenditure on conserved land, $r_j b_j$, the international agency's budget is

$$\tau \int_0^1 c_k dk = \int_0^1 (\eta w_j z_j + s r_j b_j) dj. \tag{11}$$

The international agency decides on the minimum proportion of conserved land, $\underline{b}$, for all country j:

$$b_j \geq \underline{b} \in [0, 1] \quad \text{for } j \in [0, 1]. \tag{12}$$

When this constraint is binding, the agency exercises *direct regulation.*

In order to avoid multiple equilibria, it is assumed that the countries are biased for a low tax rate:

Assumption 1 If the countries face two candidates for the international agency so that both of these offer the same level of welfare for them but with a different tax rate τ, then they vote for the one with a lower tax rate τ.

[3]This corresponds well to the institutions of the EU.

3 Countries

Country j pays political contributions R_j to the international agency. It is assumed, for simplicity, that the agency consists of civil servants, of which a constant proportion $g_j \in [0, 1]$ inhabits country j. It is then true that

$$\int_0^1 g_k dk = 1. \tag{13}$$

Thus, each country j gets a constant share g_j of total contributions

$$R = \int_0^1 R_k dk. \tag{14}$$

Without political contributions, country j earns output y_j and subsidies $\eta w_j z_j + s r_j b_j$ in terms of the consumption good. Given the consumption tax τ, this income is in terms of consumption equal to $(\eta w_j z_j + s r_j b_j)/(1+\tau)$. Noting (1), (5) and (7), the ratio of this 'legal' income relative to productivity, a^{γ_j}, is defined as follows:

$$\begin{aligned}
&(y_j + \eta w_j z_j + s r_j b_j)/\big[(1+\tau)a^{\gamma_j}\big] \\
&\quad = \big[f(l_j, n_j) + \eta z_j f_l(l_j, n_j) + s f_n(l_j, n_j) b_j\big]/(1+\tau) \\
&\quad = \big[f(l_j, 1-b_j) + (1-l_j)\eta f_l(l_j, 1-b_j) + s f_n(l_j, 1-b_j) b_j\big]/(1+\tau) \\
&\quad \doteq \phi(l_j, b_j, s, \eta, \tau).
\end{aligned} \tag{15}$$

The budget constraint of country j is given by

$$(1+\tau) p c_j = p(y_j + \eta w_j z_j + s r_j b_j) + g_j R - R_j, \tag{16}$$

where c_j is consumption, τ the consumption tax, p the price of the consumption good, $y_j + \eta w_j z_j + s r_j b_j$ the 'legal' income, R_j the contributions to the international agency and $g_j R$ the proportion of total contributions in country j. Noting (4), (15) and (16), consumption in country j is determined by

$$\begin{aligned}
c_j &= (y_j + \eta w_j z_j + s r_j b_j)/(1+\tau) + (g_j R - R_j)/\big[p(1+\tau)\big] \\
&= a^{\gamma_j} \phi(l_j, b_j, s, \eta, \tau) + (g_j R - R_j) b^{-\delta}.
\end{aligned} \tag{17}$$

Noting (3) and (17), the expected utility of country j starting at time T is

$$\begin{aligned}
\Gamma_j &= E \int_T^\infty c_j b^\delta e^{-\rho(\theta - T)} d\theta \\
&= E \int_T^\infty a^{\gamma_j} \frac{b^\delta}{1+\tau} \phi(l_j, b_j, s, \eta, \tau) e^{-\rho(\theta - T)} d\theta + \frac{g_j R - R_j}{\rho},
\end{aligned} \tag{18}$$

where E is the expectation operator. Country j maximizes (18) by labor input l_j and conserved land b_j subject to technological change (9) and the regulatory constraint

(12), taking the tax τ, the subsidies (s, η), biodiversity b, and the contributions (R_j, R) as given. This maximization and the symmetry throughout the countries $j \in [0, 1]$ imply (cf. Appendix A):

(i)

$$l_j = l, \qquad b_j = b \quad \textit{and} \quad n_j = 1 - b_j = 1 - b \quad \textit{for } j \in [0, 1], \tag{19}$$

(ii) *the equilibrium value of the function* ϕ:

$$\phi(l, b, s, \eta, \tau) = f(l, 1 - b), \tag{20}$$

(iii) *the first-order condition for conserved land* b_j:

$$(1 - s)\xi\left(\frac{l}{1 - b}\right) = \left[(1 - l)\eta - \frac{slb}{1 - b}\right]\frac{lf_{ll}(l, 1 - b)}{f(l, 1 - b)} \quad \text{for } b > \underline{b}, \tag{21}$$

(iv) *the first-order condition for labor input in production,* l_j:

$$(1 - \eta)\left[1 - \xi\left(\frac{l}{1 - b}\right)\right] + \left[(1 - l)\eta - \frac{slb}{1 - b}\right]\frac{lf_{ll}(l, 1 - b)}{f(l, 1 - b)} = \frac{(a - 1)\lambda l}{\rho + (1 - a)\lambda(1 - l)}, \tag{22}$$

(v) *the value function* Γ_j:

$$\Gamma_j(b, \gamma_j, s, \eta, \tau, R, R_j), \quad \partial\Gamma_j / \partial R_j = -1/\rho,$$

$$\frac{\partial \Gamma_j}{\partial b} = \frac{b^{\delta - 1}\Omega_j}{1 + \tau}\left[\frac{b}{1 - b}\left\{(s - 1)\xi\left(\frac{l}{1 - b}\right) + \left[(1 - l)\eta - \frac{sl}{1 - b}\right]\frac{lf_{ll}(l, 1 - b)}{f(l, 1 - b)}\right\} + \delta\right] \quad \text{for } b = \underline{b}, \tag{23}$$

$$\frac{\partial \Gamma_j}{\partial b} = \frac{b^{\delta}}{1 + \tau}\delta\frac{\Omega_j}{b} = \delta\frac{b^{\delta - 1}\Omega_j}{1 + \tau} \quad \text{for } b > \underline{b},$$

where Ω_j is the maximum value of $E\int_T^{\infty} a^{\gamma_j}\phi e^{-\rho(\theta - T)}d\theta$.

4 The Pareto Optimum

Assume a *benevolent* international agency that claims no political contributions, $R_j = 0$ for all j, uses subsidies (s, η) to both abatement investment and conserved land, and maximizes the expected value of the geometric average of the utility of the countries in the whole economy:

$$E\int_T^{\infty} cb^{\delta}e^{-\rho(\theta - T)}d\theta \quad \text{with } \log c \doteq \int_0^1 \log c_j dj. \tag{24}$$

Because the agency controls the allocation of resources completely by the subsidies (s, η), it attains the Pareto optimum (l^P, b^P) (cf. Appendix B):

$$\left[1-\xi\left(\frac{l^P}{1-b^P}\right)\right]\left[\rho+(1-a)\lambda\left(1-l^P\right)\right]=(a-1)\lambda l^P, \tag{25}$$

$$\frac{b^P}{1-b^P}\xi\left(\frac{l^P}{1-b^P}\right)=\delta. \tag{26}$$

5 Direct Regulation

Assume a *self-interested* international agency that has no budget of its own, $s = \eta = \tau = 0$, controls the proportion of conserved land directly by setting $b = \underline{b}$, and maximizes the present value of the expected flow of the political contributions at time T [cf. (13)],[4]

$$E\int_T^\infty \int_0^1 R_j e^{-\rho(\theta-T)} d\theta = \frac{1}{\rho}\int_0^1 R_j dj. \tag{27}$$

In line with Grossman and Helpman (1994), a common agency game is constructed as follows. First, the countries set their political contributions R_j conditional on the international agency's prospective policy b, taking total contributions R as given.[5] Second, the international agency sets b and collects the contributions. Third, the countries maximize their expected utility given the contributions R_j and R. The game is solved in reverse order: first for a country (stage 3) and then for the political equilibrium (stages 2 and 1).

With direct regulation, labor input l_j is the only instrument and (22) the only equilibrium condition for country j. Noting (12), the value function (23) and the equilibrium condition (22) for country j take the form

$$\Gamma_j\left(b^R, \gamma_j, 0, 0, 0, R, R_j\right), \tag{28}$$

$$(a-1)\lambda l^R = \left[\rho+(1-a)\lambda\left(1-l^R\right)\right]\left[1-\xi\left(\frac{l^R}{1-b^R}\right)\right]. \tag{29}$$

[4]This is a modification of the idea of Grossman and Helpman (1994), who assume that a policy maker's welfare is a linear function of both the political contributions and the utilities of the lobbies. This characterizes the fact that the policy maker cares about (a) its revenue from political contributions and (b) the possibility of being re-elected, which depends of the utility of the electorate (i.e. the members of the lobbies). This setup is simplified by ignoring the utilities of the lobbies. Because the policy instruments must maximize the utility of each lobby in equilibrium [cf. condition (iii) in Appendix C], the results would not change if Grossman and Helpman's original welfare function were used.

[5]The crucial point in the common agency game is that each country j can credibly commit itself to its contribution function $R_j(b)$.

The international agency maximizes the present value (27). Each country j maximizes the value of its optimal program, (28), by influencing the international agency by its contributions R_j, but taking total contributions R as given. Because b^R is a policy and $R_j(b^R)$ the strategy of country j, the equilibrium conditions of this game are [cf. (ii) and (iii) in Appendix C]

$$b = \arg\max_{b^R} \frac{1}{\rho} \int_0^1 R_j(b^R)dj = \arg\max_{b^R} \int_0^1 R_j(b^R)dj, \tag{30}$$

$$b = \arg\max_{b^R} \Gamma_j\big(b^R, \gamma_j, 0, 0, 0, R, R_j(b^R)\big) \quad \text{for } j \in [0,1]. \tag{31}$$

With (23) and $\eta = s = \tau = 0$, the condition (31) is equivalent to

$$0 = \frac{\partial \Gamma_j}{\partial b} + \frac{\partial \Gamma_j}{\partial R_j} R_j' = \big(b^R\big)^{\delta-1} \Omega_j \left[-\frac{b^R}{1-b^R} \xi\left(\frac{l^R}{1-b^R}\right) + \delta \right] - \frac{1}{\rho} R_j'$$

and

$$R_j' = \rho \big(b^R\big)^{\delta-1} \Omega_j \left[\delta - \frac{b^R}{1-b^R} \xi\left(\frac{l^R}{1-b^R}\right) \right]. \tag{32}$$

Finally, given (32), the condition (30) is equivalent to

$$0 = \frac{1}{\rho} \int_0^1 R_j' dj = \left[\delta - \frac{b^R}{1-b^R} \xi\left(\frac{l^R}{1-b^R}\right) \right] \big(b^R\big)^{\delta-1} \int_0^1 \Omega_j dj. \tag{33}$$

Equations (29) and (33) satisfy the conditions (25) and (26). This result can concluded as follows:

Proposition 1 *Direct regulation is Pareto optimal,* $(l^R, b^R) = (l^P, b^P)$.

The international agency, benevolent or self-interested, eliminates the externality due to biodiversity as a macroeconomic decision-maker.

6 Conservation Subsidies

Assume a self-interested international agency that imposes the conservation subsidy s. Assume furthermore that because the agency cannot fully distinguish between abatement investment and other labor expenditures, the abatement investment subsidy η is incentive incompatible. Without losing any generality, one can then choose $\eta = 0$.

In this common agency game, the subsidy s is a public policy instrument. With $\eta = 0$, the value function (23) and the equilibrium conditions (21) and (22) for country j become

$$\Gamma_j\big(b^S, \gamma_j, s, 0, \tau, R, R_j\big), \tag{34}$$

$$s\frac{(l^S)^2 b^S f_{ll}(l^S, 1-b^S)}{(1-b^S) f(l^S, 1-b^S)} = (s-1)\xi\left(\frac{l^S}{1-b^S}\right), \tag{35}$$

$$\frac{(a-1)\lambda l^S}{\rho+(1-a)\lambda(1-l^S)} = 1-\xi-\frac{s(l^S)^2 f_{ll}}{(1-b^S)f} = 1-s\xi\left(\frac{l^S}{1-b^S}\right). \tag{36}$$

In this setup, the budget constraint (11) becomes (cf. Appendix D)

$$\tau = \frac{sb^S}{1-b^S}\xi\left(\frac{l^S}{1-b^S}\right). \tag{37}$$

In the three equations (35), (36) and (37), there are three unknown variables τ, l^S and b^S, and one known variable s. This system defines the functions

$$\tau(s), \qquad l^S(s), \qquad b^S(s). \tag{38}$$

Unfortunately, the derivatives of these functions are mathematically ambiguous. For this reason, one can make the plausible assumption that the direct effect of the subsidy s dominates. This implies that the following holds true:

Assumption 2 An increase in the subsidy s to conserved land increases both the supply of conserved land, $(b^S)' > 0$, and the tax that is needed for financing the increase of the subsidy, $\tau' > 0$.

The international agency maximizes the present value of the expected flow of the political contributions at time T, (27). Country j maximizes the value of its optimal program, (34), by influencing the international agency by its contributions R_j, but taking total contributions R as given. Because s is a policy and $R_j(s)$ the strategy of country j, then, given (38), the equilibrium conditions are [cf. (ii) and (iii) in Appendix C]:

$$s = \arg\max_s \frac{1}{\rho}\int_0^1 R_j(s)dj = \arg\max_s \int_0^1 R_j(s)dj, \tag{39}$$

$$s = \arg\max_s \Gamma_j\big(b, \gamma_j, s, 0, \tau(s), R, R_j(s)\big) \quad \text{for } j \in [0,1]. \tag{40}$$

From (5), (8) and (35) it follows that conserved land is subsidized:

$$s = \Bigg[\underbrace{\xi}_{+} - \underbrace{\frac{(l^S)^2 b^S f_{ll}}{(1-b^S)f}}_{-}\Bigg]^{-1}\underbrace{\xi}_{+} > 0.$$

Given (37), this subsidy $s > 0$ must be financed by the wage tax $\tau > 0$. To show that $(l^S, b^S) \neq (l^P, b^P)$, assume $(l^S, b^S) = (l^P, b^P)$. In that case, relations (25), (36) and (38) lead to the following contradiction:

$$0 = 1 - \xi\left(\frac{l^P}{1-b^P}\right) - \frac{(a-1)\lambda l^P}{\rho + (1-a)\lambda(1-l^P)}$$
$$= 1 - \xi\left(\frac{l^S}{1-b^S}\right) - \frac{(a-1)\lambda l^S}{\rho + (1-a)\lambda(1-l^S)} = \frac{sl^S}{1-b^S}\frac{l^S f_{ll}(l^S, 1-b^S)}{f(l^S, 1-b^S)} \neq 0.$$

Thus, $(l^S, b^S) \neq (l^P, b^P)$ holds true. This and Proposition 1 imply that:

Proposition 2 *The equilibrium with conservation subsidies is Pareto suboptimal, $(l^S, b^S) \neq (l^P, b^P)$. Consequently, a switch from regulation to conservation subsidies decreases welfare.*

This is because the international agency imposes a distorting consumption tax τ to finance the conservation subsidy s. With direct regulation, there is no distorting taxation.

Because the equilibrium (l^S, b^S) is Pareto suboptimal, then the same welfare can be attained by two tax rates τ (with corresponding subsidies s):

- With the higher tax rate τ, the subsidy s is higher and consequently, the amount of conserved land is bigger than at Pareto optimum, $b^S > b^P$.
- With the lower tax rate τ, the subsidy s is lower and consequently, the amount of conserved land is smaller than at Pareto optimum, $b^S < b^P$.

Given Assumption 1, only the equilibrium with a lower tax rate, $b^S < b^P$, is feasible. Given this, Assumption 2 and Proposition 2, the following conclusion is made:

Proposition 3 *A switch from direct regulation into conservation subsidies decreases both the growth rate (i.e. $g^R = g^P > g^S$) and biodiversity in each country (i.e. $b^R = b^P > b^S$).*

Because any inefficiency decreases the resources of the economy, there are less resources to be put into abatement investment and the conservation of biodiversity.

7 Conclusions

This paper considers an economy in which the conservation of land yields utility through biodiversity, firms improve their efficiency by in-house abatement investment and local interest groups lobby a self-interested international agency over biodiversity management. Two policy alternatives are compared: the regulation of land use and subsidies for conserved land. The main findings are the following.

In the case of direct regulation, the international agency determines the use of land throughout the whole economy, fully internalizing the externality through biodiversity. In the case of conservation subsidies, revenue-raising taxes cause distortions. For this reason, a shift from subsidies to direct regulation increases the resources of the countries, promoting investment in abatement investment and economic growth. The transfer of labor from production to abatement investment decreases the demand for land in production. This promotes the conservation of land and biodiversity.

While a great deal of caution should be exercised when a highly stylized game-theoretical model is used to derive results on growth and biodiversity, the following conclusion seems to be justified. The prospect of lobbying changes the outcome of biodiversity management fundamentally. A larger package of policy instruments leads to Pareto improvement with a benevolent international agency, but to Pareto worsening with a self-interested one. In the case of Natura 2000, for instance, regulation without any budget may be an appropriate degree of authority for the Commission. Greater authority narrows biodiversity and slows down economic growth.

Acknowledgements The author is grateful to Timothy Swanson and Duncan Knowler for their constructive comments.

Appendix A: Equations (21) and (22) and Function (23)

Noting (5) and (8), the function (15) has the partial derivatives:

$$\begin{aligned}\frac{\partial \phi}{\partial b_j} &= (s-1) f_n(l_j, 1-b_j) - (1-l_j)\eta f_{ln}(l_j, 1-b_j) - s f_{nn}(l_j, 1-b_j) b_j \\ &= \left\{ (s-1)\xi\left(\frac{l_j}{1-b_j}\right) + \left[(1-l_j)\eta - \frac{s l_j b_j}{1-b_j}\right] \frac{l_j f_{ll}(l_j, 1-b_j)}{f(l_j, 1-b_j)} \right\} \\ &\quad \times \frac{f(l_j, 1-b_j)}{1-b_j} = 0 \quad \text{for } b > \underline{b}, \end{aligned} \tag{41}$$

$$\begin{aligned}\frac{\partial \phi}{\partial l_j} &= (1-\eta) f_l(l_j, 1-b_j) + (1-l_j)\eta f_{ll}(l_j, 1-b_j) + s f_{ln}(l_j, 1-b_j) b_j \\ &= (1-\eta)\left[1 - \xi\left(\frac{l_j}{1-b_j}\right)\right] \frac{f(l_j, 1-b_j)}{l_j} \\ &\quad + \left[(1-l_j)\eta - \frac{s l_j b_j}{1-b_j}\right] f_{ll}(l_j, 1-b_j). \end{aligned} \tag{42}$$

The maximization of the expected utility (18) by (l_j, b_j) s.t. (9) and (12), given $(\tau, s, \eta, b, R_j, R)$, is equivalent to the maximization of

$$E \int_T^\infty a^{\gamma_j} \frac{b^\delta}{1+\tau} \phi(l_j, b_j, s, \eta, \tau) e^{-\rho(\theta - T)} d\theta$$

s.t. (9) and (12), given $(\tau, s, \eta, b, R_j, R)$. The value of this optimal program starting at time T is

$$\Omega_j(\gamma_j, \underline{b}, s, \eta, \tau) \doteq \max_{(l_j, b_j) \text{ s.t. } (9), (12)} E \int_T^\infty \frac{b^\delta}{1+\tau} a^{\gamma_j} \phi(l_j, b_j, s, \eta, \tau) e^{-\rho(\theta - T)} d\theta. \quad (43)$$

The Bellman equation corresponding to the optimal program (43) is given by (cf. Dixit and Pindyck 1994)

$$\rho \Omega_j = \max_{(l_j, b_j) \text{ s.t. } (9)} \Lambda^j(l_j, b_j, \gamma_j, \underline{b}, s, \eta, \tau), \quad (44)$$

where

$$\Lambda^j(l_j, b_j, \gamma_j, \underline{b}, s, \eta, \tau) = \frac{b^\delta}{1+\tau} a^{\gamma_j} \phi(l_j, b_j, s, \eta, \tau) + \lambda(1 - l_j)\big[\Omega_j(\gamma_j + 1, \underline{b}, s, \eta, \tau) - \Omega_j(\gamma_j, \underline{b}, s, \eta, \tau)\big]. \quad (45)$$

The first-order conditions corresponding to the Bellman equation (44) and (45) are $\partial \Lambda^j / \partial l_j = 0$ and $\partial \Lambda^j / \partial b_j = 0$. To solve the dynamic program, I assume that the value of the program, Ω_j, is in fixed proportion to the instantaneous utility at the optimum:

$$\Omega_j(\gamma_j, \underline{b}, s, \eta, \tau) = \varphi_j \frac{b^\delta}{1+\tau} a^{\gamma_j} \phi\big(l_j^*, b_j, s, \eta, \tau\big) \quad \text{with } b_j = b_j^* \text{ for } b_j > \underline{b}, \quad (46)$$

where φ_j is a constant and l_j^* and b_j^* are the optimal values of l_j and b_j. From (46) it follows that

$$\Omega_j(\gamma_j + 1, \underline{b}, s, \eta, \tau) / \Omega_j(\gamma_j, \underline{b}, s, \eta, \tau) = a. \quad (47)$$

Inserting (46) and (47) into the Bellman equation (44) and (45) yields

$$1/\varphi_j = \rho + (1 - a)\lambda(1 - l_j) > 0. \quad (48)$$

Noting (41), (42), (45), (47) and (48), the first-order conditions corresponding to the maximization (44) are given by

$$\frac{1}{\Omega_j} \frac{\partial \Lambda^j}{\partial b_j} = \frac{b^\delta a^{\gamma_j}}{\Omega_j} \frac{\partial \phi}{\partial b_j} = \frac{b^\delta a^{\gamma_j}}{(1+\tau)\Omega_j} \frac{f(l_j, 1 - b_j)}{1 - b_j} \times \left\{ (s-1)\xi \left(\frac{l_j}{1 - b_j} \right) + \left[(1 - l_j)\eta - \frac{s l_j}{1 - b_j} \right] \frac{l_j f_{ll}(l_j, 1 - b_j)}{f(l_j, 1 - b_j)} \right\} = 0 \quad \text{for } b > \underline{b}, \quad (49)$$

$$
\begin{aligned}
\frac{1}{\Omega_j}\frac{\partial \Lambda^j}{\partial l_j} &= \frac{b^\delta a^{\gamma_j}}{(1+\tau)\Omega_j}\frac{\partial \phi}{\partial l_j} - \lambda\left[\frac{\Omega_j(\gamma_j+1,\underline{b},s,\eta,\tau)}{\Omega_j(\gamma_j,\underline{b},s,\eta,\tau)} - 1\right] \\
&= \frac{1}{\varphi_j \phi}\frac{\partial \phi}{\partial l_j} - (a-1)\lambda \\
&= \left[\rho + (1-a)\lambda(1-l_j)\right]\frac{1}{\phi}\frac{\partial \phi}{\partial l_j} - (a-1)\lambda \\
&= \left[\rho + (1-a)\lambda(1-l_j)\right]\frac{f(l_j,1-b_j)}{(1+\tau)\phi} \\
&\quad \times \left\{(1-\eta)\left[1-\xi\left(\frac{l_j}{1-b_j}\right)\right]\frac{1}{l_j}\right. \\
&\quad \left. + \left[(1-l_j)\eta - \frac{s l_j b_j}{1-b_j}\right]\frac{f_{ll}(l_j,1-b_j)}{f(l_j,1-b_j)}\right\} - (a-1)\lambda \\
&= 0. \qquad (50)
\end{aligned}
$$

In the system consisting of the international agency budget (11) and the first-order conditions (49) and (50) for all countries $j \in [0,1]$, there is symmetry throughout $j \in [0,1]$. This implies $l_j = l$ and $b_j = b$ for $j \in [0,1]$. From this, (1), (5), (6), (13), (14), (15) and (17) it follows that

$$
\begin{aligned}
\phi(l,b,s,\eta,\tau) &= \int_0^1 a^{\gamma_j}\phi(l,b,s,\eta,\tau)dj \Big/ \int_0^1 a^{\gamma_k}dk \\
&= \Bigg[\int_0^1 a^{\gamma_j}\phi(l,b,s,\eta,\tau)dj + b^{-\delta}\underbrace{\int_0^1 (g_j R - R_j)dj}_{=0}\Bigg] \Big/ \int_0^1 a^{\gamma_k}dk \\
&= \int_0^1 \left[a^{\gamma_j}\phi(l,b,s,\eta,\tau) + (g_j R - R_j)b^{-\delta}\right]dj \Big/ \int_0^1 a^{\gamma_k}dk \\
&= \int_0^1 c_j dj \Big/ \int_0^1 a^{\gamma_k}dk \\
&= \int_0^1 y_j dj \Big/ \int_0^1 a^{\gamma_k}dk \\
&= f(l,1-b). \qquad (51)
\end{aligned}
$$

This implies (20). Inserting (51), $l_j = l$ and $b_j = b$ back to (49) and (50) yields (21) and (22).

Noting $l_j = l$, $b_j = b$, (15), (43), (46) and (51), the expected utility of country j, (18), can be written as follows:

$$
\Gamma_j(b,\gamma_j,\underline{b},s,\eta,\tau,R,R_j) \doteq \Omega_j(\gamma_j,\underline{b},s,\eta,\tau) + (g_j R - R_j)/\rho, \qquad \frac{\partial \Gamma_j}{\partial R_j} = -\frac{1}{\rho},
$$

$$\left.\frac{\partial \Gamma_j}{\partial \underline{b}}\right|_{b=\underline{b}} = \frac{b^\delta}{1+\tau}\left[\frac{\partial \Omega_j}{\partial b_j} + \delta\frac{\Omega_j}{b}\right] = \frac{b^\delta}{1+\tau}\left[\frac{\Omega_j}{\phi}\frac{\partial \phi}{\partial b_j} + \delta\frac{\Omega_j}{b}\right]$$

$$= \frac{b^\delta}{1+\tau}$$

$$\times\left\{\frac{\Omega_j}{\phi}\left[(s-1)f_n - (1-l_j)\eta f_{ln} - s f_{nn}\right] + \delta\frac{\Omega_j}{b}\right\}$$

$$= \frac{b^\delta \Omega_j}{1+\tau}$$

$$\times\left[\frac{1}{\phi}\left\{(s-1)\frac{f(l_j, 1-b_j)}{1-b_j}\xi\right.\right.$$

$$\left.\left. + \left[(1-l)\eta - \frac{sl}{1-b}\right]\frac{l f_{ll}}{1-b}\right\} + \frac{\delta}{b}\right]$$

$$= \frac{b^\delta \Omega_j}{1+\tau}\left[\frac{f(l, 1-b)}{(1-b)\phi(l, b, s, \eta, \tau)}\right.$$

$$\left.\times\left\{(s-1)\xi + \left[(1-l)\eta - \frac{slb}{1-b}\right]\frac{l f_{ll}}{f}\right\} + \frac{\delta}{b}\right]$$

$$= \frac{b^{\delta-1}\Omega_j}{1+\tau}\left[\frac{b}{1-b}\left\{(s-1)\xi\left(\frac{l}{1-b}\right)\right.\right.$$

$$\left.\left. + \left[(1-l)\eta - \frac{slb}{1-b}\right]\frac{l f_{ll}(l, 1-b)}{f(l, 1-b)}\right\} + \delta\right],$$

$$\frac{\partial \Gamma_j}{\partial b} = \frac{b^\delta}{1+\tau}\delta\frac{\Omega_j}{b} = \delta\frac{b^{\delta-1}\Omega_j}{1+\tau} \quad \text{for } b > \underline{b}.$$

Appendix B: Equations (25) and (26)

The average serial number of technology in the economy is given by

$$\gamma = \int_0^1 \gamma_j dj. \tag{52}$$

Given the Poisson property of the improvement of technology in countries $j \in [0, 1]$ (cf. Sect. 2.2), one obtains the following. In a small period of time dt, the probability that abatement investment will lead a jump from γ to $\gamma + 1$ is given by $\lambda z dt$, while the probability that abatement investment will remain without success is given by $1 - \lambda z dt$. Noting (9), this defines a Poisson process χ with

$$d\chi = \begin{cases} 1 & \text{with probability } \lambda(1-l)dt, \\ 0 & \text{with probability } 1 - \lambda(1-l)dt, \end{cases} \qquad l \doteq \int_0^1 l_j dj, \tag{53}$$

where $d\chi$ is the increment of the process χ.

Because there is perfect symmetry throughout countries $j \in [0,1]$ in the system (2), (53), (21) and (22), there is $l_j = l = l^P$ and $b_j = b = b^P$ for $j \in [0,1]$ in equilibrium. Because there is one-to-one correspondence from (η, s) to (l^P, b^P), one can replace the subsidies (η, s) by (l^P, b^P) as the international agency's policy instruments. Thus, the international agency maximizes (24) by (l^P, b^P) s.t. technological change (53). Noting (5), (17) and (52), one obtains the value function of this maximization as follows:

$$\Delta(l^P, b^P) \doteq E\int_T^\infty cb^\delta e^{-\rho(\theta-T)}d\theta = f\big(l^P, 1-b^P\big)\big(b^P\big)^\delta E\int_T^\infty a^\gamma e^{-\rho(\theta-T)}d\theta$$
$$= \frac{f(l^P, 1-b^P)(b^P)^\delta}{\rho + (1-a)\lambda l^P}.$$

Noting (8), this leads to the first-order conditions

$$\frac{\partial \log \Delta}{\partial l^P} = \frac{f_l(l^P, 1-b^P)}{f(l^P, 1-b^P)} + \frac{(1-a)\lambda}{\rho + (1-a)\lambda l^P}$$
$$= \frac{1}{l^P}\left[1 - \xi\left(\frac{l^P}{b^P}\right)\right] + \frac{(1-a)\lambda}{\rho + (1-a)\lambda l^P} = 0,$$
$$\frac{\partial \log \Delta}{\partial l^P} = \frac{\delta}{b^P} - \frac{f_m(l^P, 1-b^P)}{f(l^P, 1-b^P)} = \frac{\delta}{b^P} - \frac{1}{1-b^P}\xi\left(\frac{l^P}{b^P}\right) = 0.$$

These equations imply (25) and (26).

Appendix C: The Lobbying Game

Following Dixit et al. (1997), a subgame perfect Nash equilibrium for this game is a policy ζ and a set of contribution schedules $R_1(\zeta), \ldots, R_J(\zeta)$ such that the following conditions (i)–(iv) hold:

(i) Contributions R_j are non-negative but no more than the contributor's income, $\Gamma_j \geq 0$.
(ii) The policy ζ maximizes the international agency's welfare (27) taking the contribution schedules R_j as given.
(iii) Country j cannot have a viable strategy $R_j(\zeta)$ that yields it a higher level of utility than in equilibrium, given the others' contributions.
(iv) Country j provides the international agency at least with the level of utility as in the case in which it offers nothing ($R_j = 0$), and the international agency responds optimally given the contribution functions of the other countries.

Appendix D: Equation (37)

Given $\eta = 0$, (7), (8), (13), (14), (17), (19) and (20), the international agency budget constraint (11) becomes

$$\tau = \frac{\int_0^1 (\eta w_j z_j + s r_j b_j) dj}{\int_0^1 c_k dk} = \frac{s \int_0^1 r_j b_j dj}{\int_0^1 c_k dk} = \frac{s \int_0^1 a^{\gamma_j} f_n(l_j, n_j) b_j dj}{\int_0^1 c_k dk}$$

$$= \frac{s \int_0^1 a^{\gamma_j} f_n(l_j, n_j) b_j dj}{\int_0^1 a^{\gamma_k} \phi(l_k, b_k, s, \eta, \tau) dk + \underbrace{\int_0^1 (g_k R - R_k) b^{-\delta} dk}_{=0}} = \frac{s \int_0^1 a^{\gamma_j} f_n(l_j, n_j) b_j dj}{\int_0^1 a^{\gamma_k} \phi(l_k, b_k, s, \eta, \tau) dk}$$

$$= s \frac{f_n(l^S, 1 - b^S) b^S}{\phi(l^S, b^S, s, 0, \tau)} = s \frac{f_n(l^S, 1 - b^S) b^S}{f(l^S, 1 - b^S)} = \frac{s b^S}{1 - b^S} \xi \left(\frac{l^S}{1 - b^S} \right).$$

References

Aghion, P., & Howitt, P. (1998). *Endogenous growth theory*. Cambridge: MIT Press.

Barbier, E. B., & Schulz, C.-E. (1997). Wildlife, biodiversity and trade. *Environment and Development Economics*, *2*(2), 145–172.

Dixit, A., & Pindyck, K. (1994). *Investment under uncertainty*. Princeton: Princeton University Press.

Dixit, A., Grossman, G. M., & Helpman, E. (1997). Common agency and coordination: general theory and application to management policy making. *Journal of Political Economy*, *105*(4), 752–769.

Endres, A., & Radke, V. (1999). Land use, biodiversity, and sustainability. *Journal of Economics*, *70*(1), 1–16.

Grossman, G., & Helpman, E. (1994). Protection for sale. *American Economic Review*, *84*(4), 833–850.

Johal, S., & Ulph, A. (2002). Globalization, lobbying, and international environmental governance. *Review of International Economics*, *10*(3), 387–403.

Jordan, A. J. (1998). Step change or stasis? EC environmental policy after the Amsterdam summit. *Environmental Politics*, *7*(1), 227–236.

MacArthur, R., & Wilson, E. O. (1967). *The theory of island biography*. Princeton: Princeton University Press.

Majone, G. (1996). *Regulating Europe*. London: Routledge.

Ledoux, L., Crooks, S., Jordan, A., & Turner, K. (2000). Implementing EU biodiversity policy: UK experiences. *Land Use Policy*, *17*(4), 257–268.

Ostermann, O. P. (1998). The need for management of nature conservation sites designated under Natura 2000. *Journal of Applied Ecology*, *35*(6), 968–973.

Rowthorn, B., & Brown, G. (1999). When a high discount rate encourages biodiversity. *International Economic Review*, *40*(2), 315–332.

Swanson, T. M. (1994). The economics of extinction revisited and revised: a generalized framework for the analysis of the problems of endangered species and biodiversity losses. *Oxford Economics Papers*, *46*(4), 800–821.

Weber, N., & Christophersen, T. (2002). The influence of non-governmental organizations on the creation of Natura 2000 during the European policy process. *Forest policy and economics*, *4*(1), 1–12.

Environmental Regulations, Abatement and Economic Growth

Elke Moser, Alexia Prskawetz, and Gernot Tragler

1 Introduction

In recent years climate change and the possible consequences that human society might have to deal with, if further global warming cannot be stopped, have become one of the most important topics in science, politics and the world wide media. The scientific evidence that many key climate indicators are already moving beyond the patterns of natural variability defines this dramatic change as a world wide concern. Hence, the importance of climate mitigation has become undeniable. These indicators, including global mean surface temperature, global ocean temperature, global average sea level, northern hemisphere snow cover and Arctic sea ice decline as well as extreme climatic events, additionally come along with the risk of abrupt or irreversible climatic shifts, which might have devastating consequences for the entire world population. This underlines how urgent the need of climate actions has become (see Richardson et al. 2009).

In the 4th Assessment Report by the International Panel on Climate Change (IPCC 2007), scientific evidence on global warming, its damages and the importance of climate mitigation as well as the reduction of anthropogenic greenhouse

E. Moser (✉) · A. Prskawetz
Wittgenstein Centre (IIASA, VID/ÖAW, WU), VID/ÖAW, Wohllebengasse 12-14, Vienna 1040, Austria
e-mail: elke.moser@oeaw.ac.at

A. Prskawetz
Research Group Economics, Institute of Mathematical Methods in Economics, Vienna University of Technology, Argentinierstrasse 8, Vienna 1040, Austria
e-mail: afp@econ.tuwien.ac.at

G. Tragler
Research Group ORCOS, Institute of Mathematical Methods in Economics, Vienna University of Technology, Argentinierstrasse 8, Vienna 1040, Austria
e-mail: gernot.tragler@tuwien.ac.at

J. Crespo Cuaresma et al. (eds.), *Green Growth and Sustainable Development*, Dynamic Modeling and Econometrics in Economics and Finance 14,
DOI 10.1007/978-3-642-34354-4_5, © Springer-Verlag Berlin Heidelberg 2013

gas (GHG) emissions are highlighted. According to their *Synthesis Report*, the industry sector, besides the energy supply and transport sectors, is one of the main sources of anthropogenic GHG emissions accounting for almost 20% of all GHG emissions (2004). The majority are CO_2 emissions due to the use of fossil fuels, but also the emissions of other gases like PFCs, SF_6, CH_4 and N_2O due to physical and chemical processes contribute to the overall CO_2 emissions. Additionally, one has to consider the impact of industrial waste and wastewater on pollution. Further on, not only the sources are discussed in the IPCC (2007) but also a broad range of mitigation policy measures are suggested, which especially emphasizes the role of technology policies and the increasing need for more R&D efforts. In the *Mitigation of Climate Change Report*, some possible mitigation options for a greener technology are explained, such as fuel switching, including the use of waste material, advanced energy efficiency, the use of bioenergy and material recycling and substitution. As far as according policy instruments are concerned, they consider performance standards, subsidies, tax credits, tradeable permits and voluntary agreements as the most environmentally effective instruments.

Although these environmental policy instruments seem to be promising, the question arises how they can be utilized in the most effective way and whether strict environmental regulation has a supporting or repressing impact on innovation and economic growth. To address this issue, many economic growth models include the environment as an additional dimension in form of pollution that is modeled either as a by-product of production like in Kalkuhl et al. (2011) or Saltari and Travaglini (2011), or as a result of consumption, as in Bretschger and Smulders (2007). To reduce pollution in order to protect the environment, the possibility of end-of-pipe abatement often is added in such models, like in Lange and Moslener (2004), Rasmussen (2001), Antweiler et al. (2001) or Xepapadeas (1992). Instead of reducing pollution only after production, a different approach is to reduce pollution directly in the process of production by including a cleaner substitute for the pollutive production input or for the pollutive technology. Examples for this can be found in Cunha-e Sá et al. (2010), Hartley et al. (2010), Cassou and Hamilton (2004), Acemoglu et al. (2009) and Lehmijoki and Palokangas (2010).

We refer in our work to a recent paper by Rauscher (2009) who addresses this topic by constructing a simple dynamic environmental-economic model which considers capital accumulation, end-of-pipe emission abatement, R&D investments and knowledge spillovers in an endogenous growth framework. Rauscher investigates in a conveniently tractable way whether tighter environmental standards will induce a shift from end-of-pipe emission abatement to a process-integrated one and how these alternative policies effect R&D investments and growth. The model Rauscher employs is kept algebraically simple without specifying concrete functional forms. In this paper we introduce specific functional forms and apply optimal control theory to solve for the dynamic paths of the environmental-economic system.

The paper is organized as follows. In the next section we introduce the model, which is solved Sect. 3 by applying Pontryagin's Maximum Principle. Numerical simulations, including a bifurcation analysis, are presented in Sect. 4. Section 5 concludes and gives an outlook for further research.

2 The Model

To investigate the effects of environmental standards on economic growth and R&D investments, we build on the model by Rauscher (2009), who considers a competitive market economy where a continuum of identical firms using identical technologies produce a homogenous GDP good. In this economy two types of capital are accumulated: first, there is *conventional capital*, also called *brown capital*, which is pollutive, secondly, a non-polluting *green capital* can be chosen. Additionally, the government sets environmental standards which the entrepreneurs are obligated to meet. The necessary abatement effort as well as the abatement costs depend on the stringency of these regulations. Consequently, firms adopting cleaner technologies have to spend less on end-of-pipe abatement. This benefit, however, comes at a cost because the required resources for green R&D could be invested otherwise profitably in conventional R&D. Instead of assuming different groups of agents, as frequently done in many other papers approaching this topic, Rauscher (2009) focuses on one type of agent in the private sector of the economy, who is a capital-owning entrepreneur doing his/her R&D in-house and who saves and consumes all at the same time. In case of perfect competition of the markets on which these agents interact, the simple homogenous-representative-agent model generates the same results as its more elaborated version with heterogeneous agents.

Maximizing his/her own profit, the representative agent has to consider the present value of future utility, given as

$$\int_0^\infty e^{-rt}\big(\ln\big(C(t)\big)+u(\varepsilon)\big)dt \quad \text{with } C(t)>0, \tag{1}$$

where $C(t)$ is the consumption or dividend income, $\ln(C(t))$ describes the utility level that the agent obtains from $C(t)$ and r is the discount rate. Further on, ε specifies the exogenously given environmental quality determined by the government, which is represented by index between 0 and 1, with $\varepsilon = 0$ denoting the laissez-faire scenario (any environmental regulation exists and therefore environmental quality is low) and $\varepsilon = 1$ stands for the maximal attainable environmental quality. The private sector's utility of environmental quality is denoted as $u(\varepsilon)$ and will be set in the following as $u(\varepsilon) = c\varepsilon^\gamma$ with $c > 0$ and $0 < \gamma < 1$.

The entrepreneurs use conventional capital $K(t)$ and/or green capital $G(t)$ to produce an output

$$F\big(K(t), G(t)\big) = bK(t)^{\alpha_1}G(t)^{\alpha_2}$$
$$\text{with } b>0,\ t\in[0,\infty),\ 0<\alpha_2\le\alpha_1<1 \text{ and } \alpha_1+\alpha_2\le 1. \tag{2}$$

Output is used for consumption, for the coverage of opportunity costs due to green and brown R&D investments and for end-of-pipe emission abatement. Note, that savings are not included in this model approach. The budget constraint is given as follows,

$$F\big(K(t), G(t)\big) - C(t) - w\big(R_K(t) + R_G(t)\big) - \chi(\varepsilon)K(t) = 0. \tag{3}$$

Note that as of here, we will often omit the time argument t for the ease of exposition. R_K and R_G denote the investments for R&D to generate new capital of types K and G, respectively. The parameter $w \in [0, 1]$ represents the exogenous opportunity costs. The abatement costs for achieving the binding environment constraints of the government are proportional to the installed conventional capital K. The costs per unit capital is given as $\chi(\varepsilon)$ which is increasing and convex in the stringency of environmental regulation, i.e. $\chi' > 0$, $\chi'' > 0$, and will be set for this analysis as $\chi(\varepsilon) = a\varepsilon^{\beta}$ with $a > 0$ and $\beta > 1$.

The two types of capital accumulate through a Cobb Douglas production function with decreasing returns to scale and depreciate at fixed exogenous rates ϕ and ψ,

$$\dot{K} = A(K, R_K) = dK^{\delta_1} R_K^{\delta_2} - \phi K \quad \text{with } \delta_1 + \delta_2 < 1, \tag{4}$$

$$\dot{G} = B(G, R_G) = eG^{\sigma_1} R_G^{\sigma_2} - \psi G \quad \text{with } \sigma_1 + \sigma_2 < 1. \tag{5}$$

The existing capital stock itself has a positive feedback on the accumulation. Assuming that this positive feedback is weaker than the contribution of new technology due to R&D, the partial elasticity of production of the capital stock is supposed to be less than the one of the R&D investments. Hence, $\delta_1 < \delta_2$ and $\sigma_1 < \sigma_2$. Additionally, it is more likely that conventional capital is more established in the economy than green one and therefore accumulation is much easier. To take this imbalance into account, the partial elasticities of green capital G should at least not be greater than those of conventional capital K, i.e. $\sigma_1 \leq \delta_1$ and $\sigma_2 \leq \delta_2$.

Figure 1 shows the interrelations of the variables to illustrate the dynamics of the model. Starting from the capital stocks K and G, output $F(K, G)$ is produced. Constricted by the available budget (3), the decision-maker has to determine the extend of R&D investments that are made for either brown (R_K) or green (R_G) capital or possibly both. These investments in turn influence the growth of the capital stocks K and G, respectively. Additionally, also the existing capital stock contributes to the accumulation.

Solving (3) for consumption C together with (1) leads to an optimal control problem with R_K and R_G as control variables and the two available types of capital as states, which is given as

$$\max_{R_K, R_G} \int_0^{\infty} e^{-rt}\Big(\ln\big(bK^{\alpha_1}G^{\alpha_2} - w(R_K + R_G) - a\varepsilon^{\beta}K\big) + c\varepsilon^{\gamma}\Big)dt \tag{6a}$$

$$\text{s.t.:} \quad \dot{K} = dK^{\delta_1} R_K^{\delta_2} - \phi K \tag{6b}$$

$$\dot{G} = eG^{\sigma_1} R_G^{\sigma_2} - \psi G \tag{6c}$$

$$0 \leq R_K \quad \forall t \geq 0 \tag{6d}$$

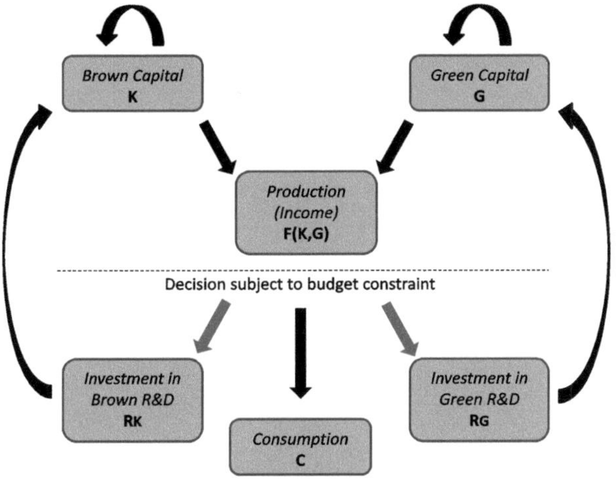

Fig. 1 Sketch of the dynamics of the model

$$0 \leq R_G \quad \forall t \geq 0 \tag{6e}$$

$$0 < bK^{\alpha_1}G^{\alpha_2} - w(R_K + R_G) - a\varepsilon^{\beta}K \tag{6f}$$

$$0 \leq \varepsilon \leq 1 \tag{6g}$$

$$0 < \alpha_1, \alpha_2, \gamma, w < 1 \quad \text{and} \quad \alpha_1 + \alpha_2 \leq 1 \tag{6h}$$

$$0 < \delta_1, \delta_2 < 1 \quad \text{and} \quad \delta_1 + \delta_2 < 1 \tag{6i}$$

$$0 < \sigma_1, \sigma_2 < 1 \quad \text{and} \quad \sigma_1 + \sigma_2 < 1 \tag{6j}$$

$$1 < \beta \tag{6k}$$

$$0 < \phi, \psi, a, b, c, d, e, r. \tag{6l}$$

3 Analytical Results

3.1 Derivation of the Canonical System

Summing up, we consider a discounted autonomous model with infinite planning horizon. To derive the necessary conditions for an optimal solution we consider the Lagrangian $\mathcal{L}$ in current value notation, where $\mathcal{H}$ denotes the Hamiltonian, $\mathcal{C}$ the control and mixed path constraints and μ the vector of Lagrange Multipliers:

$$\begin{aligned}
\mathcal{L} &= \mathcal{H} + \mu\mathcal{C} \\
&= \lambda_0\big(\ln\big(F(K,G) - w(R_K + R_G) - \chi(\varepsilon)K\big) + u(\varepsilon)\big) \\
&\quad + \lambda_1 A(K, R_K) + \lambda_2 B(G, R_G) + \mu_1 R_K + \mu_2 R_G \\
&\quad + \mu_3\big(F(K,G) - w(R_K + R_G) - \chi(\varepsilon)K\big)
\end{aligned}$$

with the co-states $(\lambda_0, \lambda_1, \lambda_2) \neq 0$. The first order conditions are

$$\mathcal{L}_{R_K} = \frac{-w\lambda_0}{F(K,G) - w(R_K + R_G) - \chi(\varepsilon)K} + \lambda_1 A_{R_K} + \mu_1 - w\mu_3 = 0 \quad (7)$$

$$\mathcal{L}_{R_G} = \frac{-w\lambda_0}{F(K,G) - w(R_K + R_G) - \chi(\varepsilon)K} + \lambda_2 B_{R_G} + \mu_2 - w\mu_3 = 0 \quad (8)$$

$$\dot{\lambda_1} = \lambda_1(r - A_K) - \lambda_0 \frac{F_K(K,G) - \chi(\varepsilon)}{F(K,G) - w(R_K + R_G) - \chi(\varepsilon)K} - \mu_3\big(F_K(K,G) - \chi(\varepsilon)\big) \quad (9)$$

$$\dot{\lambda_2} = \lambda_2(r - B_G) - \lambda_0 \frac{F_G(K,G)}{F(K,G) - w(R_K + R_G) - \chi(\varepsilon)K} - \mu_3 F_G(K,G) \quad (10)$$

where subscripts denote partial derivatives of multivariate functions. The complementary slackness conditions are

$$\begin{aligned} &\mu_1 \geq 0 \quad \text{and} \quad 0 = \mu_1 R_K \\ &\mu_2 \geq 0 \quad \text{and} \quad 0 = \mu_2 R_G \\ &\mu_3 \geq 0 \quad \text{and} \quad 0 = \mu_3\big(F(K,G) - w(R_K + R_G) - \chi(\varepsilon)K\big). \end{aligned} \quad (11)$$

One can show that $\lambda_0 = 1$, without loss of generality. For the derivation of the canonical system one has to distinguish between the different cases of an interior arc and a boundary arc. In the first case none of the constraints are active and, due to the complementary slackness conditions in (11), $(\mu_1, \mu_2, \mu_3) = 0$. Hence, an optimal control should maximize the current value Hamiltonian, i.e.

$$\big(R_K^*, R_G^*\big) = \arg \max_{(R_K, R_G)} \mathcal{H}$$

and therefore

$$\mathcal{L}_{R_K} = \mathcal{H}_{R_K} = 0 \quad (12)$$

$$\mathcal{L}_{R_G} = \mathcal{H}_{R_G} = 0. \quad (13)$$

To prove that the Hamiltonian is strictly concave, the positivity of the co-states is necessary which can be shown by solving (12) and (13) for λ_1 and λ_2 respectively. This yields

$$\lambda_1 = \frac{w}{(F(K,G) - w(R_K + R_G) - a\varepsilon^\beta K) A_{R_K}(K, R_K)} > 0$$

$$\lambda_2 = \frac{w}{(F(K,G) - w(R_K + R_G) - a\varepsilon^\beta K) B_{R_G}(G, R_G)} > 0.$$

Note that $A_{R_K R_K}(K, R_K) < 0$ and $B_{R_G R_G}(G, R_G) < 0$. The Hessian matrix of the Hamiltonian

$$H = \begin{pmatrix} -\frac{w^2}{(F(K,G)-w(R_K+R_G)-\chi(\varepsilon)K)^2} + \lambda_1 A_{R_K R_K}(K, R_K) & -\frac{w^2}{(F(K,G)-w(R_K+R_G)-\chi(\varepsilon)K)^2} \\ -\frac{w^2}{(F(K,G)-w(R_K+R_G)-\chi(\varepsilon)K)^2} & -\frac{w^2}{(F(K,G)-w(R_K+R_G)-\chi(\varepsilon)K)^2} + \lambda_2 B_{R_G R_G}(G, R_G) \end{pmatrix}$$

therefore is negative definite and the Hamiltonian $\mathcal{H}$ is strictly concave.

The optimality conditions in (12) and (13) allow to derive control functions depending on co-state and state variables (cf. conditions (7) and (8))

$$\begin{aligned} R_K &= R_K(K, G, \lambda_1, \lambda_2) \\ R_G &= R_G(K, G, \lambda_1, \lambda_2). \end{aligned} \tag{14}$$

Substituting these control functions into the state dynamics (4) and (5) as well as into the adjoint equations (9) and (10) the canonical system in the state-co-state-space is given as

$$\begin{aligned} \dot{K} &= A\big(K, R_K(K, G, \lambda_1, \lambda_2)\big) \\ \dot{G} &= B\big(G, R_G(K, G, \lambda_1, \lambda_2)\big) \\ \dot{\lambda_1} &= \lambda_1(r - A_K) \\ &\quad - \frac{F_K(K, G) - \chi(\varepsilon)}{F(K, G) - w(R_K(K, G, \lambda_1, \lambda_2) + R_G(K, G, \lambda_1, \lambda_2)) - \chi(\varepsilon)K} \\ \dot{\lambda_2} &= \lambda_2(r - B_G) \\ &\quad - \frac{F_G(K, G)}{F(K, G) - w(R_K(K, G, \lambda_1, \lambda_2) + R_G(K, G, \lambda_1, \lambda_2)) - \chi(\varepsilon)K}. \end{aligned}$$

However, from an application orientated point of view it is often more convenient to transform the canonical system from the state-co-state-space into the state-control-space. Within this representation immediate interpretation of the results is more convenient (see Grass et al. 2008). Additionally, inserting the specific functions from above, the two controls from (7) and (8) are given only implicitly. Therefore, the derivation of the canonical system in the state-control space is even necessary.

Considering the specific functions from above, the first order conditions are

$$\begin{aligned} \mathcal{H}_{R_K} &= -\frac{w}{bK^{\alpha_1}G^{\alpha_2} - w(R_K + R_G) - a\varepsilon^{\beta}K} + \lambda_1\big(dK^{\delta_1}\delta_2 R_K^{\delta_2 - 1}\big) \\ &= 0 \end{aligned} \tag{15a}$$

$$\begin{aligned} \mathcal{H}_{R_G} &= -\frac{w}{bK^{\alpha_1}G^{\alpha_2} - w(R_K + R_G) - a\varepsilon^{\beta}K} + \lambda_2\big(eG^{\sigma_1}\sigma_2 R_G^{\sigma_2 - 1}\big) \\ &= 0 \end{aligned} \tag{15b}$$

$$\dot{\lambda_1} = \lambda_1\left(r - d\delta_1 K^{\delta_1 - 1} R_K^{\delta_2} + \phi\right) - \frac{\alpha_1 b K^{\alpha_1 - 1} G^{\alpha_2} - a\varepsilon^{\beta}}{b K^{\alpha_1} G^{\alpha_2} - w(R_K + R_G) - a\varepsilon^{\beta} K} \tag{15c}$$

$$\dot{\lambda_2} = \lambda_2\left(r - e\sigma_1 G^{\sigma_1 - 1} R_G^{\sigma_2} + \psi\right) - \frac{\alpha_2 b K^{\alpha_1} G^{\alpha_2 - 1}}{b K^{\alpha_1} G^{\alpha_2} - w(R_K + R_G) - a\varepsilon^{\beta} K}. \tag{15d}$$

Solving (15a) and (15b) for λ_1 and λ_2 instead of the controls yields

$$\begin{aligned} \lambda_1(K, G, R_K, R_G) &= \frac{w}{(b K^{\alpha_1} G^{\alpha_2} - w(R_K + R_G) - a\varepsilon^{\beta} K) d K^{\delta_1} \delta_2 R_K^{\delta_2 - 1}} \\ \lambda_2(K, G, R_K, R_G) &= \frac{w}{(b K^{\alpha_1} G^{\alpha_2} - w(R_K + R_G) - a\varepsilon^{\beta} K) e G^{\sigma_1} \sigma_2 R_G^{\sigma_2 - 1}}. \end{aligned} \tag{16}$$

By using the total time derivatives of the co-states

$$\begin{aligned} \dot{\lambda_1} &= \lambda_{1_K} \dot{K} + \lambda_{1_G} \dot{G} + \lambda_{1_{R_K}} \dot{R_K} + \lambda_{1_{R_G}} \dot{R_G} \\ \dot{\lambda_2} &= \lambda_{2_K} \dot{K} + \lambda_{2_G} \dot{G} + \lambda_{2_{R_K}} \dot{R_K} + \lambda_{2_{R_G}} \dot{R_G} \end{aligned} \tag{17}$$

two equations for the control dynamics can be obtained. Together with the adjoint dynamics in (15c) and (15d) these control dynamics are given as

$$\begin{aligned} \dot{R_K} = &-\big(\dot{\lambda_2}\lambda_{1_{R_G}} - \dot{\lambda_1}\lambda_{2_{R_G}} \\ &+ \dot{G}(\lambda_{1_G}\lambda_{2_{R_G}} - \lambda_{1_{R_G}}\lambda_{2_G}) + \dot{K}(\lambda_{1_K}\lambda_{2_{R_G}} - \lambda_{1_{R_G}}\lambda_{2_K})\big) \\ &/(\lambda_{1_{R_K}}\lambda_{2_{R_G}} - \lambda_{1_{R_G}}\lambda_{2_{R_K}}) \\ \dot{R_G} = &-\big(\dot{\lambda_1}\lambda_{2_{R_K}} - \dot{\lambda_2}\lambda_{1_{R_K}} \\ &+ \dot{G}(\lambda_{1_{R_K}}\lambda_{2_G} - \lambda_{1_G}\lambda_{2_{R_K}}) + \dot{K}(\lambda_{1_{R_K}}\lambda_{2_K} - \lambda_{1_K}\lambda_{2_{R_K}})\big) \\ &/(\lambda_{1_{R_K}}\lambda_{2_{R_G}} - \lambda_{1_{R_G}}\lambda_{2_{R_K}}) \end{aligned} \tag{18}$$

which yields the canonical system

$$\begin{aligned} \dot{R_K} = &\, D_1^2 D_2^2 R_G^2 R_K^2 Y^3 \\ &/\big(w^2\big(d(\delta_2 - 1)\delta_2 K^{\delta_1} R_K^{\delta_2}\big(D_2 R_G^2 w - eY(\sigma_2 - 1)\sigma_2 G^{\sigma_1} R_G^{\sigma_2}\big) \\ &+ D_1 e R_K^2 w(\sigma_2 - 1)\sigma_2 G^{\sigma_1} R_G^{\sigma_2}\big)\big) \\ &\times \Big\{\big[\big(eY(\sigma_2 - 1)\sigma_2 G^{\sigma_1} R_G^{\sigma_2 - 2} - D_2 w\big) \\ &\times \big(D_1\big(a\varepsilon^{\beta} - b\alpha_1 G^{\alpha_2} K^{\alpha_1 - 1}\big) + w\big(-d\delta_1 K^{\delta_1 - 1} R_K^{\delta_2} + r + \phi\big)\big) \end{aligned}$$

$$
\begin{aligned}
&\quad + D_2 w\big(w\big(-e\sigma_1 G^{\sigma_1-1} R_G^{\sigma_2} + r + \psi\big) - bD_2\alpha_2 G^{\alpha_2-1} K^{\alpha_2}\big)\big] \\
&\quad \times \frac{w}{D_1 D_2^2 Y^3} + \dot{G} T_1 + \dot{K} T_2 \Big\} \\
\dot{R_G} &= D_1^2 D_2^2 R_G^2 R_K^2 Y^3 \\
&\quad /\big(w^2\big(d(\delta_2-1)\delta_2 K^{\delta_1} R_K^{\delta_2}\big(D_2 R_G^2 w - eY(\sigma_2-1)\sigma_2 G^{\sigma_1} R_G^{\sigma_2}\big) \\
&\quad + D_1 e R_K^2 w(\sigma_2-1)\sigma_2 G^{\sigma_1} R_G^{\sigma_2}\big)\big) \\
&\quad \times \Big\{\big[\big(dY(\delta_2-1)\delta_2 K^{\delta_1} R_K^{\delta_2-2} - D_1 w\big) \\
&\quad \times \big(w\big(-e\sigma_1 G^{\sigma_1-1} R_G^{\sigma_2} + r + \psi\big) - bD_2\alpha_2 G^{\alpha_2-1} K^{\alpha_2}\big) \\
&\quad + D_1 w\big(aD_1\varepsilon^{\beta} - bD_1\alpha_1 G^{\alpha_2} K^{\alpha_1-1} - dw\delta_1 K^{\delta_1-1} R_K^{\delta_2} \\
&\quad + w(r+\phi)\big)\big] \\
&\quad \times \frac{w}{D_1 D_2^2 Y^3} + \dot{G} T_3 + \dot{K} T_4 \Big\} \qquad (19) \\
\dot{K} &= dK^{\delta_1} R_K^{\delta_2} - \phi K \\
\dot{G} &= eG^{\sigma_1} R_G^{\sigma_2} - \psi G
\end{aligned}
$$

with

$$
\begin{aligned}
T_1 &= \frac{ew^2\sigma_2 G^{\sigma_1-1} R_G^{\sigma_2-2}(b\alpha_2(\sigma_2-1)G^{\alpha_2}K^{\alpha_1} + R_G w\sigma_1)}{D_1 D_2^2 Y^3} \\
T_2 &= -\frac{w^2}{D_1^2 D_2^2 K R_G^2 R_K Y^3} \\
&\quad \times \big(d\delta_1\delta_2 K^{\delta_1} R_K^{\delta_2}\big(D_2 R_G^2 w - eY(\sigma_2-1)\sigma_2 G^{\sigma_1} R_G^{\sigma_2}\big) \\
&\quad - D_1 e R_K(\sigma_2-1)\sigma_2 G^{\sigma_1} R_G^{\sigma_2}\big(b\alpha_1 G^{\alpha_2} K^{\alpha_1} - aK\varepsilon^{\beta}\big)\big) \\
T_3 &= \frac{w^2}{D_1^2 D_2^2 G R_G R_K^2 Y^3} \\
&\quad \times \big(d(\delta_2-1)\delta_2 K^{\delta_1} R_K^{\delta_2}\big(bD_2 R_G \alpha_2 G^{\alpha_2} K^{\alpha_1} + eY\sigma_1\sigma_2 G^{\sigma_1} R_G^{\sigma_2}\big) \\
&\quad - D_1 e R_K^2 w\sigma_1\sigma_2 G^{\sigma 1} R_G^{\sigma_2}\big) \\
T_4 &= \frac{dw^2\delta_2 K^{\delta_1-1} R_K^{\delta_2-2}((\delta_2-1)(b\alpha_1 G^{\alpha_2}K^{\alpha_1} - aK\varepsilon^{\beta}) + R_K w\delta_1)}{D_1^2 D_2 Y^3} \\
Y &= bK^{\alpha_1} G^{\alpha_2} - w(R_K + R_G) - a\varepsilon^{\beta} K
\end{aligned}
$$

and D_1 and D_2 being the first derivatives of the state dynamics with respect to the corresponding control

$$D_1 = dK^{\delta_1}\delta_2 R_K^{\delta_2 - 1}$$
$$D_2 = eG^{\sigma_1}\sigma_2 R_G^{\sigma_2 - 1}.$$

In the boundary arc case, the optimal controls do not necessarily maximize the Hamiltonian, i.e. $\mathcal{H}_{R_K} = 0$ and $\mathcal{H}_{R_G} = 0$ might not be fulfilled in the optimum. Hence, the approach to derive the canonical system in the state-control-space, as done in (15a)–(18), cannot be used. Instead, the optimal controls have to maximize the Lagrangian. Therefore, in case of one or even both control constraints being active, the partial derivatives of the Lagrange function with respect to the controls, $\mathcal{L}_{R_K} = 0$ and $\mathcal{L}_{R_G} = 0$, together with the active constraint equations yield the corresponding Lagrange multipliers and the control dynamics, while the adjoint equations can be used to calculate the co-states. The state dynamics remain the same just with the according control values inserted, i.e. $R_K = 0$ and/or $R_G = 0$. If, however, the mixed path constraint is fulfilled, the derivation of the according canonical system is more extensive. Assuming that the mixed path constraint is the only constraint being active, meaning that R_K and R_G are positive, the following DAEs have to be solved

$$\begin{aligned}
\dot{K} &= A(K, G, R_K, R_G) \\
\dot{G} &= B(K, G, R_K, R_G) \\
\dot{\lambda}_1 &= \lambda_1(r - A_K) - \frac{F_K(K, G) - \chi(\varepsilon)}{F(K, G) - w(R_K + R_G) - \chi(\varepsilon)K} \\
&\quad - \mu_3\big(F_K(K, G) - \chi(\varepsilon)\big) \\
\dot{\lambda}_2 &= \lambda_2(r - B_G) - \frac{F_G(K, G)}{F(K, G) - w(R_K + R_G) - \chi(\varepsilon)K} - \mu_3 F_G(K, G) \\
\mathcal{L}_{R_K} &= \mathcal{H}_{R_K} + \mu_3 C_{R_K} = 0 \\
\mathcal{L}_{R_G} &= \mathcal{H}_{R_G} + \mu_3 C_{R_G} = 0 \\
0 &= C(K, G, R_K, R_G)
\end{aligned}$$

where C defines the mixed path constraint and this time $\mu_3 \geq 0$. In order to transform these DAEs into ordinary differential equations (ODEs), total time derivatives have to be considered:

$$\begin{aligned}
\frac{d}{dt}\mathcal{L}_{R_K} &= (\mathcal{H}_{R_K K} + \mu_3 C_{R_K K})\dot{K} + (\mathcal{H}_{R_K G} + \mu_3 C_{R_K G})\dot{G} \\
&\quad + (\mathcal{H}_{R_K R_K} + \mu_3 C_{R_K R_K})\dot{R_K} + (\mathcal{H}_{R_K R_G} + \mu_3 C_{R_K R_G})\dot{R_G} \\
&\quad + \dot{\lambda}_1 \mathcal{H}_{R_K \lambda_1} + \dot{\lambda}_2 \mathcal{H}_{R_K \lambda_2} + \dot{\mu}_3 C_{R_K} \\
&= 0
\end{aligned}$$

$$
\begin{aligned}
\frac{d}{dt}\mathcal{L}_{R_G} &= (\mathcal{H}_{R_G K} + \mu_3 C_{R_G K})\dot{K} + (\mathcal{H}_{R_G G} + \mu_3 C_{R_G G})\dot{G} \\
&\quad + (\mathcal{H}_{R_G R_K} + \mu_3 C_{R_G R_K})\dot{R_K} + (\mathcal{H}_{R_G R_G} + \mu_3 C_{R_G R_G})\dot{R_G} \\
&\quad + \dot{\lambda_1}\mathcal{H}_{R_G \lambda_1} + \dot{\lambda_2}\mathcal{H}_{R_G \lambda_2} + \dot{\mu_3} C_{R_G} \\
&= 0 \\
\frac{d}{dt}C &= C_K \dot{K} + C_G \dot{G} + C_{R_K}\dot{R_K} + C_{R_G}\dot{R_G} = 0.
\end{aligned}
\tag{20}
$$

Inserting the according equations for $\dot{K}$, $\dot{G}$, $\dot{\lambda_1}$ and $\dot{\lambda_2}$ and solving the previous equations for $\dot{R_K}$, $\dot{R_G}$ and $\dot{\mu_3}$ yields the equations for the controls. Note, however, that $\dot{\lambda_1}$ and $\dot{\lambda_2}$ include λ_1 and λ_2 respectively, and therefore also $\dot{R_K}$, $\dot{R_G}$ are both dependent on the co-state. For this reason the reduction of the canonical system to four dimensions is not possible anymore and one has to consider all six dimensions which are given as follows

$$
\begin{aligned}
\dot{K} &= A(K, G, R_K, R_G) \\
\dot{G} &= B(K, G, R_K, R_G) \\
\dot{\lambda_1} &= r\lambda_1 - T_K - \lambda_1 A_K - \frac{T_{R_K} + \lambda_1 A_{R_K}}{w}\big(F_K - \chi(\varepsilon)\big) \\
\dot{\lambda_2} &= r\lambda_2 - T_G - \lambda_2 B_G - \frac{T_{R_G} + \lambda_2 B_{R_G}}{w} F_G \\
\dot{R_K} &= Y(K, G, R_K, R_G, \lambda_1, \lambda_2) \\
\dot{R_G} &= V(K, G, R_K, R_G, \lambda_1, \lambda_2)
\end{aligned}
\tag{21}
$$

where T denotes the target function

$$
T = \ln\big(F(K, G) - w(R_K + R_G) - \chi(\varepsilon)K\big) + u(\varepsilon)
$$

and Y and V denote the obtained results for the control dynamics, which we omit here because they are very complex and don't allow any immediate insights.

3.2 Steady States

According to the maximum principle (see Grass et al. 2008), in the following the maximization problem (6a) subject to (6b)–(6l) will be solved by determining the stable manifolds arising from the canonical system which has been derived in the previous section. The steady states of the canonical system are determined by solving $\dot{K} = 0$, $\dot{G} = 0$, $\dot{R_K} = 0$, $\dot{R_G} = 0$ simultaneously. Considering the two state dynamics, the according roots are:

$$\begin{aligned} K_{\dot{K}} &= \left(\frac{\phi}{dR_K^{\delta_2}}\right)^{1/(\delta_1-1)} \\ R_{K_{\dot{K}}} &= \left(\frac{\phi}{dK^{\delta_1-1}}\right)^{1/\delta_2} \\ G_{\dot{G}} &= \left(\frac{\psi}{eR_G^{\sigma_2}}\right)^{1/(\sigma_1-1)} \\ R_{G_{\dot{G}}} &= \left(\frac{\psi}{eG^{\sigma_1-1}}\right)^{1/\sigma_2} \end{aligned} \tag{22}$$

where subscripts denote the equation which is set to zero, respectively. Further on, also $K=0$ and $G=0$ would obviously be solutions. However, K and G occur in the denominator of $\dot{R_K}$ and $\dot{R_G}$ multiplicatively. Hence, for $K=G=0$ we find no feasible steady state solution of the canonical system. But since the intention of environmental policy is not to completely shut down the production, the main focus of this paper lies on the determination of steady states with a positive production output. Inserting the roots in (22) together with parameter values into $\dot{R_K}$ and $\dot{R_G}$, the intersection of the isoclines $\dot{R_K}=0$ and $\dot{R_G}=0$ determines the steady states. In this first approach only one steady state can be identified, which will be demonstrated in what follows.

3.3 Stability

To determine the stability of this steady state, the Jacobian matrix is used, which is given by

$$J = \begin{pmatrix} \dot{K}_K & 0 & \dot{K}_{R_K} & 0 \\ 0 & \dot{G}_G & 0 & \dot{G}_{R_G} \\ \dot{R}_{K_K} & \dot{R}_{K_G} & \dot{R}_{K_{R_K}} & \dot{R}_{K_{R_G}} \\ \dot{R}_{G_K} & \dot{R}_{G_G} & \dot{R}_{G_{R_K}} & \dot{R}_{G_{R_G}} \end{pmatrix}, \tag{23}$$

where subscripts denote partial derivatives again. Hence the characteristic polynomial is

$$\begin{aligned} P(\mu) = &\left(\dot{K}_{R_K}\dot{R}_{K_K} - (\dot{K}_K - \mu)(\dot{R}_{K_{R_K}} - \mu)\right) \\ &\times \left(\dot{G}_{R_G}\dot{R}_{G_G} - (\dot{G}_G - \mu)(\dot{R}_{G_{R_G}} - \mu)\right), \end{aligned} \tag{24}$$

Table 1 Possible cases of stability

det(J)		Discriminant	Eigenvalues (EV)	Signs of real part of EV	Behavior
>0	$Z_1, Z_2 > 0$	$X_1, X_2 > 0$	Real with opposite signs	$(+,-,+,-)$	Saddle point
	$Z_1, Z_2 < 0$	$X_1, X_2 > 0$	Real with same signs	$(-,-,+,+)$	Saddle point
		$X_1, X_2 < 0$	Complex	$(-,-,-,-)$	Stable
		$\text{sgn}(X_1) \neq \text{sgn}(X_2)$	Real and complex	$(+,+,+,+)$	Repelling
<0	$Z_1 > 0, Z_2 < 0$	$X_1, X_2 > 0$	Real	$(+,+,+,-)$	Unstable
		$X_1 < 0, X_2 > 0$	Real and complex		
	$Z_1 < 0, Z_2 > 0$	$X_1, X_2 > 0$	Real	$(-,-,-,+)$	
		$X_1 > 0, X_2 < 0$	Real and complex		

which determines four eigenvalues

$$\mu_{1,2} = \frac{\dot{K}_K + \dot{R}_{K_{R_K}}}{2} \pm \sqrt{\underbrace{\frac{(\dot{K}_K - \dot{R}_{K_{R_K}})^2}{4} + \dot{K}_{R_K}\dot{R}_{K_K}}_{X_1}}$$
$$\mu_{3,4} = \frac{\dot{G}_G + \dot{R}_{G_{R_G}}}{2} \pm \sqrt{\underbrace{\frac{(\dot{G}_G - \dot{R}_{G_{R_G}})^2}{4} + \dot{G}_{R_G}\dot{R}_{G_G}}_{X_2}}. \tag{25}$$

Considering the sign of the determinant

$$\det J = \underbrace{(\dot{K}_{R_K}\dot{R}_{K_K} - \dot{K}_K\dot{R}_{K_{R_K}})}_{:=Z_1}\underbrace{(\dot{G}_{R_G}\dot{R}_{G_G} - \dot{G}_G\dot{R}_{G_{R_G}})}_{:=Z_2},$$

the various cases summarized in Table 1 can be distinguished.

3.4 The Laissez-Faire Scenario and the Introduction of Environmental Policy

For the numerical analysis we set the parameter values as summarized in Table 2. At first, an economy is considered in which no environmental standards at all are imposed, i.e. $\varepsilon = 0$. In this laissez-faire scenario, the agent does not have to fulfill any environmental restrictions and therefore is completely free of abatement costs. However, this comes at the expense of environmental quality and consequently of the utility it yields. Anyway, as long as the utility of consumption is high enough to compensate for the loss of environmental quality, the agent's capital accumulation is conceivable. Due to the fact that green capital is less productive than brown capital

Table 2 Parameter values

Parameter	Value	Description
a	1	Constant of proportionality of abatement costs
b	1	Scale parameter of the production function
c	5	Scale parameter describing the utility of environmental quality
d	1	Scale parameter of $\dot{K}$
e	1	Scale parameter of $\dot{G}$
r	0.05	Discount rate
w	0.1	Opportunity cost of research
β	2	Exponent of abatement costs
γ	0.4	Exponent describing the utility of environmental quality
δ_1	0.3	Production elasticity of K in $\dot{K}$
δ_2	0.5	Production elasticity of R_K in $\dot{K}$
σ_1	0.3	Production elasticity of G in $\dot{G}$
σ_2	0.4	Production elasticity of R_G in $\dot{G}$
ϕ	0.05	Depreciation rate of $\dot{K}$
ψ	0.05	Depreciation rate of $\dot{G}$

it is obvious that the agent will mainly use the polluting capital as much as possible. However, complete abandonment of green capital is not possible due to the assumption of a Cobb Douglas production function, but the green input factor is expected to be comparatively low. Figure 2 shows that the single steady state is at $K = 29{,}160$, $G = 4{,}126$ with control levels $R_K = 4{,}453$ and $R_G = 1{,}187$, which is a saddle point according to the first case in Table 1. Obviously K is dominant in production. The colored region in Fig. 2 corresponds to the admissible region according to the mixed path constraint $C \geq 0$.

In the next step, an economy with a medium environmental quality standard $\varepsilon = 0.4$ is considered. As one can see in Fig. 3, this causes a big change in the position of the steady state. In this scenario, the saddle point is at $K = 714$, $G = 981$, $R_K = 24$ and $R_G = 96$. Due to the higher abatement costs, brown capital as dominant input factor has become too expensive. Green capital now is an essential substitute, despite its lower productivity. Comparing Fig. 3 with Fig. 2 one can see that the admissible region $C \geq 0$ shrinks with increasing ε.

Figure 4 finally shows the steady state for the basic model with constant returns to scale (CRS) in the production function, which is at $K = 904{,}808$, $G = 104{,}374$, $R_K = 545{,}908$ and $R_G = 333{,}154$. One can see that these equilibrium values are quite high, compared to the previous two scenarios. Also the admissible region expands with constant returns instead of decreasing returns to scale.

Fig. 2 Steady state in the laissez-faire scenario for $\alpha_1 = 0.6$ and $\alpha_2 = 0.2$

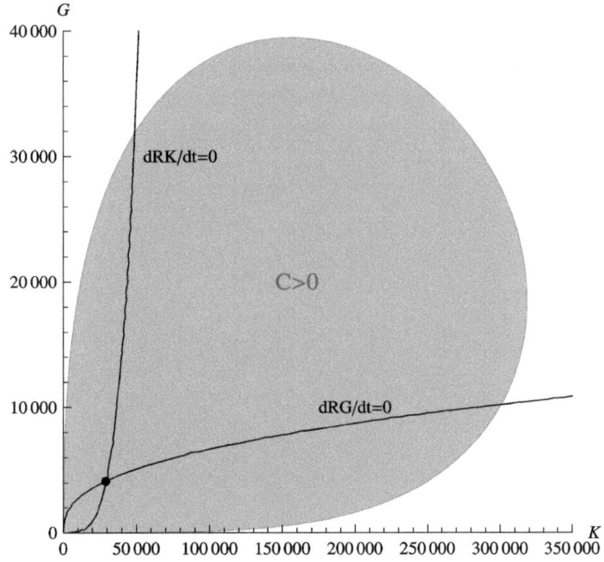

Fig. 3 Steady state for $\alpha_1 = 0.6$, $\alpha_2 = 0.2$ and $\varepsilon = 0.4$

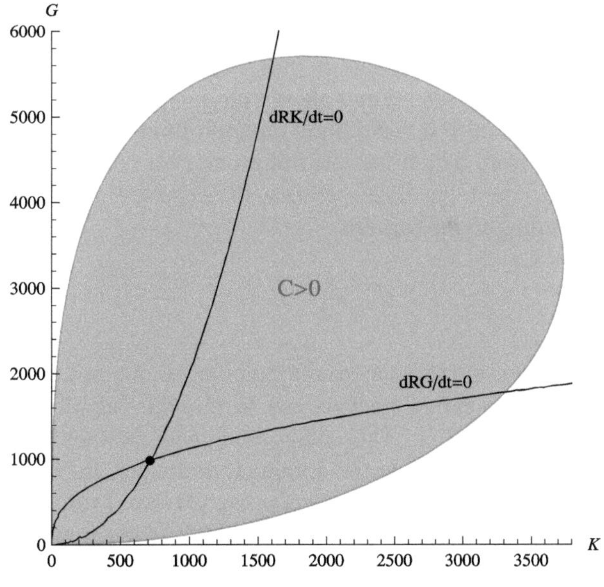

4 Optimal Paths

In this section, the matter of interest is to find trajectories converging toward the equilibrium and to get the corresponding projections that cover a significant part of the (K, G)-plane. For this purpose, the initial value problem approach is used. Hence, initial values for a backward solution of the four-dimensional canonical system need to be constructed first. However, note that only the stable manifold leads

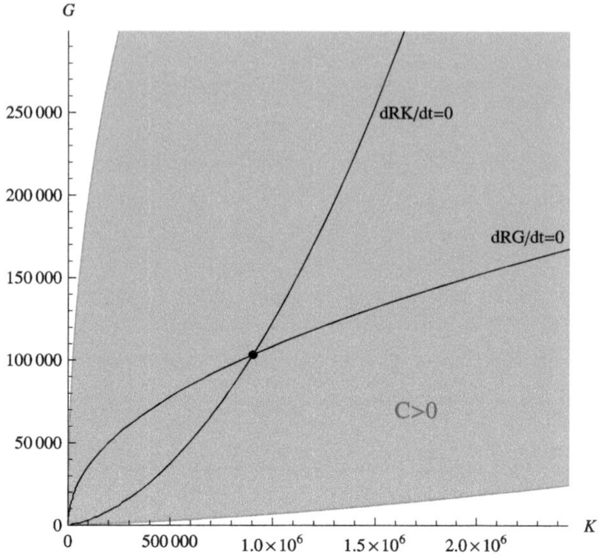

Fig. 4 Steady state for CRS with $\alpha_1 = 0.7$, $\alpha_2 = 0.3$ and $\varepsilon = 0.4$

directly into the equilibrium. Consequently, this set of starting points has to be very close to the equilibrium, in order to stay on or at least close to the stable manifold. Additionally, also dominant directions in the convergence to the steady state have to be considered. Therefore, an appropriate ellipse around the equilibrium is generated from which these starting points are taken. To take the dominant directions into account, the eigenvectors with negative eigenvalues are used for the calculation according to the formula

$$S = E + \frac{e_1}{|e_1|}\cos(\eta) + \frac{e_2}{|e_2|}\sin(\eta) \quad \text{with } \eta \in [0, 2\pi], \tag{26}$$

where S is the calculated starting point, E denotes the equilibrium, and e_1 and e_2 are the corresponding eigenvectors. Within this calculation the values of the angle η are close to $\frac{\pi}{2}$ and $\frac{3\pi}{2}$. This comes along with the fact that in those cases $\cos(\eta)$ is close to zero and therefore the dominant directions are weighted less here (cf. Knoll and Zuba 2004). Based on these constructed initial values the canonical system is solved backward. The projection of the resulting four-dimensional optimal trajectories onto the (K, G)-plane leads to a phase portrait, from which those trajectories have to be chosen, which correspond to the given initial conditions. In Fig. 5 the phase portrait for $\varepsilon = 0.4$ is depicted. Here, the crucial and obviously very narrow intervals for the angle η are $[0.4999755\pi, 0.4999756\pi]$ and $[1.500024418\pi, 1.500024419\pi]$.

As one can see in Fig. 5, some of the trajectories are divided into two parts. The first part, which is common for all and depicted in gray, corresponds to the backward solution of the system starting from the equilibrium. On the left hand side the trajectories are continued until $K = 0$. On the right hand side, however, continuation aborts when the trajectories reach the boundary of the admissible region subject to

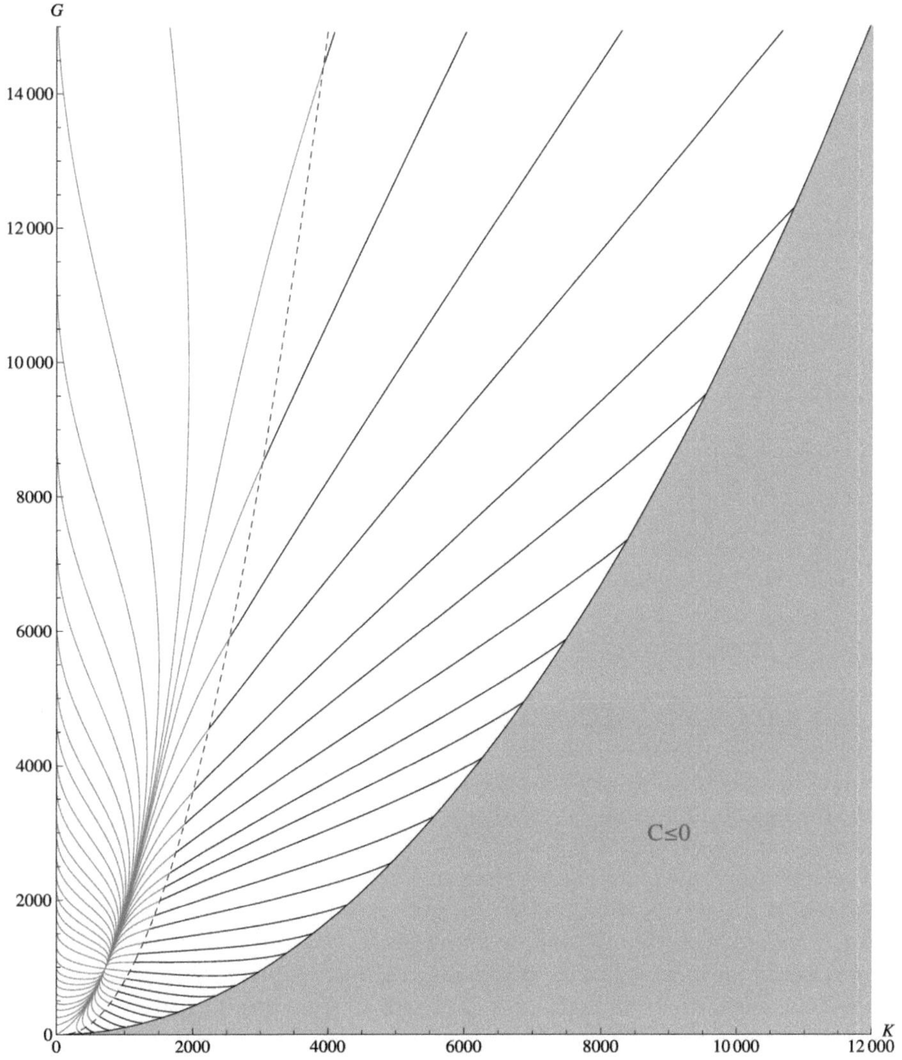

Fig. 5 Phase portrait in (K, G)-space for $\alpha_1 = 0.6$, $\alpha_2 = 0.2$ and $\varepsilon = 0.4$

the control constraint in (6d) where $R_K = 0$. This constraint is depicted in the figure as dashed black line. To enable further continuation of these trajectory paths, R_K is constantly set to zero and calculation continues with the according canonical system where $\dot{R_K} = 0$. These second parts of the trajectories are depicted in black and their continuation is possible until they finally reach the admissible boundary of the mixed path constraint in (6f), where consumption, and therefore also utility from consumption, is zero.

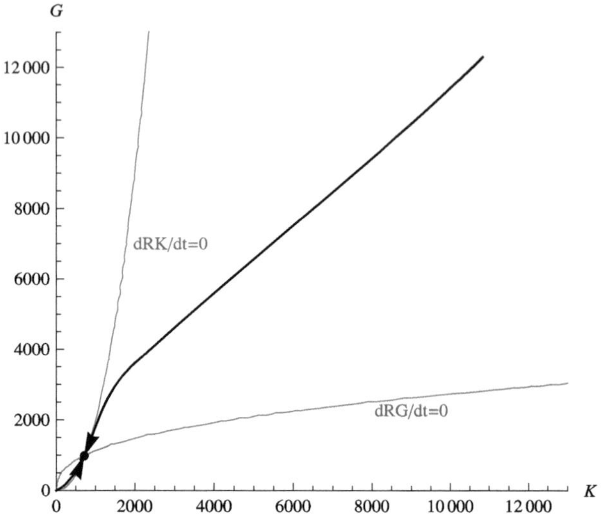

Fig. 6 Two trajectories for $\alpha_1 = 0.6$, $\alpha_2 = 0.2$ and $\varepsilon = 0.4$ with equal initial capital levels

4.1 Initial Points with an Equal Level of K and G

Figure 6 shows two trajectories from the phase portrait in the (K, G)-plane which both have initial points with almost equal levels of K and G. The first one starts at very low levels of brown and green capital which are smaller than the equilibrium values. Along the path to the equilibrium the levels of both types of capital increase. The second trajectory has its initial point at a high level of brown and green capital above the equilibrium values. Accordingly, the levels of capital decrease along the trajectory while approaching the equilibrium.

Figure 7 shows the optimal time paths in K, G, R_K and R_G along the trajectory starting at the lower level of capital. As one can see, the levels of both types of capital increase monotonously while converging toward their equilibrium values, where conventional capital in the beginning is a little bit higher than green capital. Nevertheless, green capital finally gets dominant. Considering the paths of the R&D investments, the levels of R_K and R_G initially increase very quickly. Therefore less

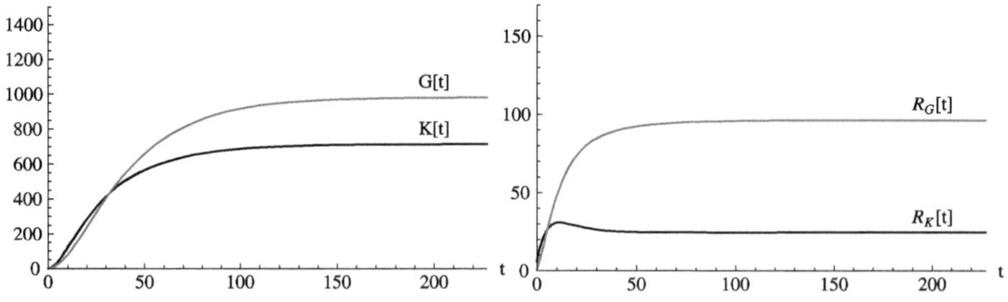

Fig. 7 Optimal time paths of state and control starting from low capital levels for $\alpha_1 = 0.6$, $\alpha_2 = 0.2$ and $\varepsilon = 0.4$

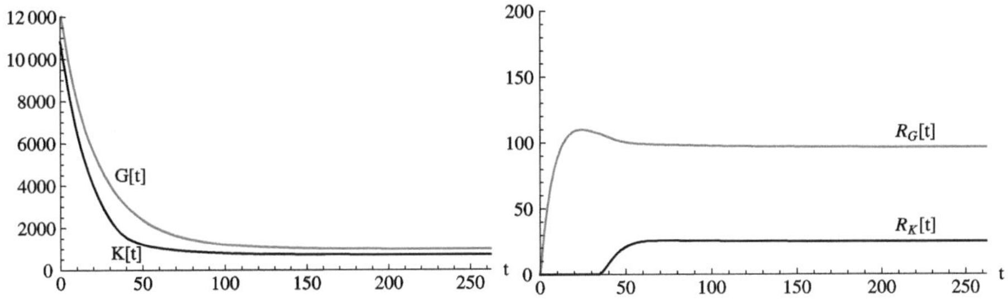

Fig. 8 Optimal time paths of state and control starting from high capital levels for $\alpha_1 = 0.6$, $\alpha_2 = 0.2$ and $\varepsilon = 0.4$

time is needed to get close to their equilibrium values. In order to cause growth in the capital levels, initially high R&D investments are needed until the positive feedback of the capital stock on itself is effective enough to thwart the negative pressure of depreciation. Note that the level of R_K even decreases after reaching a peak to slow down this positive feedback until growth and depreciation are perfectly balanced close to the equilibrium. Due to the fact that the production elasticity of R_G is less than the one of R_K, the behavior is different here. Higher investments are necessary to achieve the same effects and the R_G level monotonously increases toward the equilibrium value.

In Fig. 8 the same paths are considered for the trajectory starting at the high capital level. Here the levels of both capitals are decreasing. Due to the almost equal initial level of K and G and the comparatively lower equilibrium level of K, the decline of K is stronger than in green capital. To switch off the positive feedback of K on its own stock completely, and therefore to boost the negative impact of depreciation, R_K initially is even zero and only rises again to stop this decline, but stays at a very low level, though. Due to lower production elasticity the level of green R&D initially rises very quickly up to a peak to stop the negative pressure of depreciation. Then it slightly decreases again to finally remain at a level obviously higher than the one of R_K.

4.2 Initial Points with One Type of Capital Being Dominant

As mentioned above the initial use of both capital types is assumed due to the use of a Cobb Douglas production function. However, situations in which one type of capital is definitely the dominant input factor, whereas the other one almost equals zero, are certainly of interest. Figure 9 shows two trajectories for such initial conditions. One either starts at a green capital-dominated production or in an initial point where K is used almost exclusively as production input. In both cases, the level of the dominant capital lies above the equilibrium values, while the level of the dominated capital is below its equilibrium level.

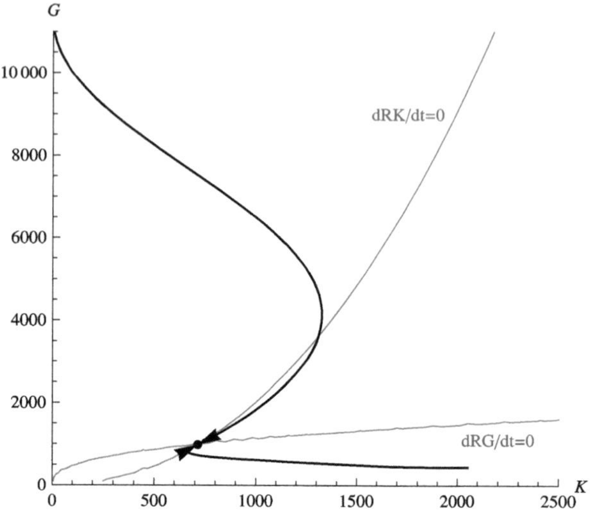

Fig. 9 Two trajectories starting at a one-capital-type-dominated production for $\alpha_1 = 0.6$, $\alpha_2 = 0.2$ and $\varepsilon = 0.4$

Figure 10 shows the optimal time paths in the case of an initially green capital-dominated production. In contrast to the previous case of an almost balanced initial mix of production, the behavior of the capital levels in this scenario are respectively opposed. Because green capital is dominant here, the level of G decreases while brown capital, starting at a very low level, rises up to the equilibrium value. Considering the R&D investments, the same behavior as in Fig. 7 can be observed, where R_K rises up to a peak, then falls again and slows down the positive feedback, while R_G increases monotonously. Summarizing this scenario it is interesting to see that R_G is increasing while G is decreasing. In other words, green R&D investments are made so to keep G at a sufficiently high level.

Regarding the case of an initially brown capital-dominated production, the according optimal time paths are depicted in Fig. 11. Accordingly, in this case K decreases and G rises up to the equilibrium values. Again, R_K is initially zero and rises up to slow down the decline, while R_G rises up to a peak and then slightly decreases.

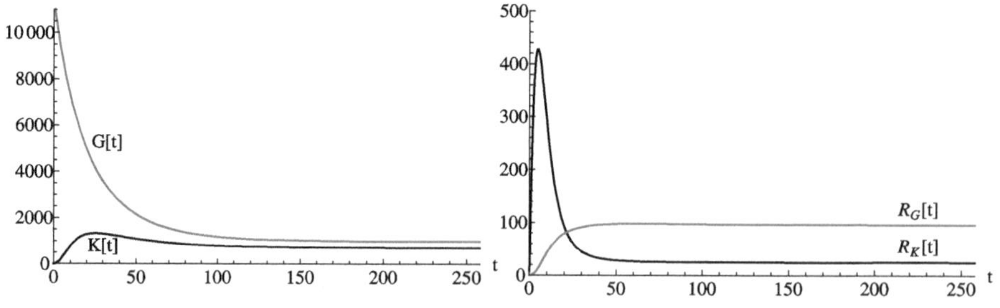

Fig. 10 Optimal time paths of state and control starting from a definitely green capital-dominated production for $\alpha_1 = 0.6$, $\alpha_2 = 0.2$ and $\varepsilon = 0.4$

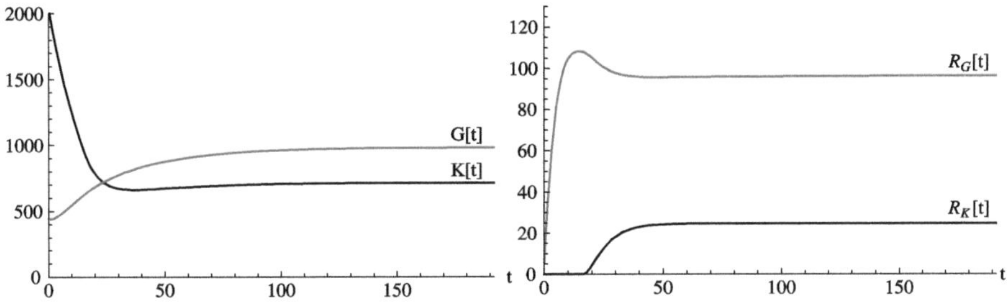

Fig. 11 Optimal time paths of state and control starting from a definitely brown capital-dominated production for $\alpha_1 = 0.6$, $\alpha_2 = 0.2$ and $\varepsilon = 0.4$

Fig. 12 Bifurcation diagram for steady state levels of K and G with respect to ε for $\alpha_1 = 0.6$, $\alpha_2 = 0.2$

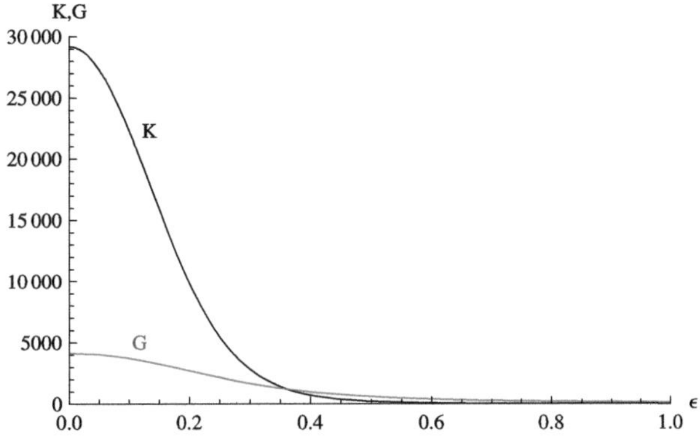

4.3 Bifurcation Analysis

In the previous sections, equilibria for specific values of ε were considered. However, the main focus of this paper is the investigation of the influence of the required environmental standards on the capital accumulation and hence on the production. We therefore apply bifurcation analysis is used with ε being the parameter to be varied. Although only one steady state has been detected so far, and hence the bifurcation diagram for the basic model is quite simple, it gives a first idea about the interrelation of the environmental quality and the usage of both types of capital as input in production.

Figure 12 depicts the change of the equilibrium values under the variation of the environmental quality imposed by the government. For $\varepsilon = 0$ (laissez-faire scenario) K is clearly dominant in production as already mentioned above. As one can see, increasing ε results in an immediate decrease of K due to the rising abatement costs per unit of brown capital. Also G decreases with growing environmental quality. This might seem a little bit astonishing at first sight, but comes along with the fact that, due to the Cobb Douglas production function, a complete abandonment of K as production input is impossible, and therefore a sufficiently small level of K has to be used which at the same time has an increasingly absorbing impact on the productivity of G. However, this decrease is much smaller than the one of K. The

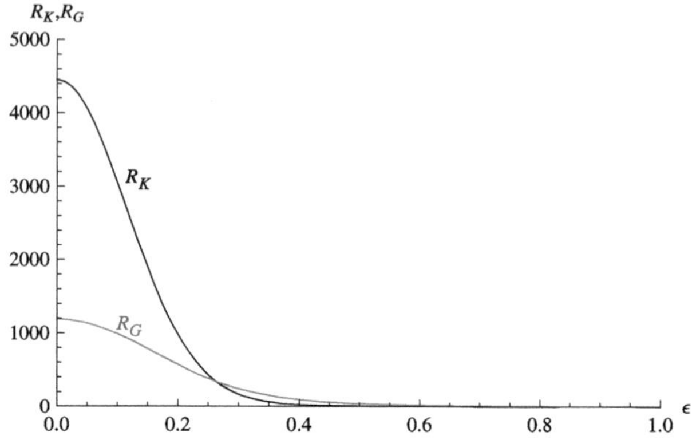

Fig. 13 Bifurcation diagram for steady state levels of R_K and R_G with respect to ε for $\alpha_1 = 0.6, \alpha_2 = 0.2$

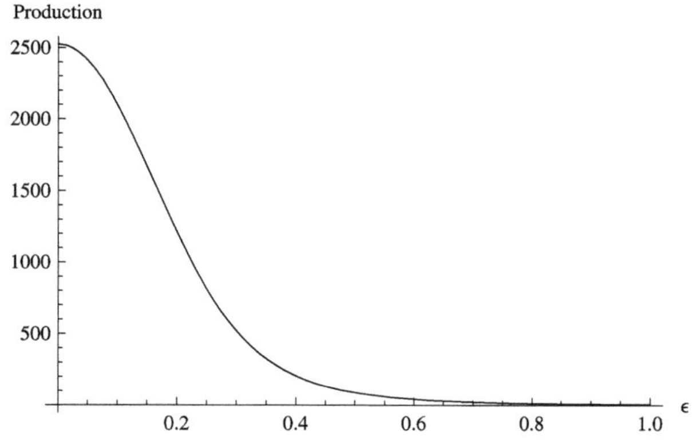

Fig. 14 Bifurcation diagram of the steady state production output with respect to ε for $\alpha_1 = 0.6, \alpha_2 = 0.2$

point of special interest is at $\varepsilon = 0.362$. At this point, abatement gets so expensive that the use of green capital as dominant production input is more advantageous. In Fig. 13 changes of the equilibrium values of R_K and R_G over ε are shown. They behave quite similarly. Initially, R_K is dominant until abatement gets too expensive and higher investments in green R&D are optimal. This change happens already at $\varepsilon = 0.263$, i.e. earlier than for the capital stocks.

Note, however, that in this basic model increasing environmental standards in general have a diminishing impact on the production inputs, and therefore on production output, and furthermore on economic growth. As one can see in Fig. 14, the production is strictly monotonously decreasing.

In contrast, the utility function as depicted in Fig. 15 rises up to a peak before it decreases due to the trade-off between consumption and environmental quality. If ε is small enough, a small loss in consumption in return for a slightly better environment is advantageous. The utility-maximizing environmental quality is at $\varepsilon = 0.125$.

In order to get a more qualitative comparison of the changing use of K and G in production with increasing ε, the percentage values of green and brown capital in total production are shown in Fig. 16. As one can see, the ratio of G follows a

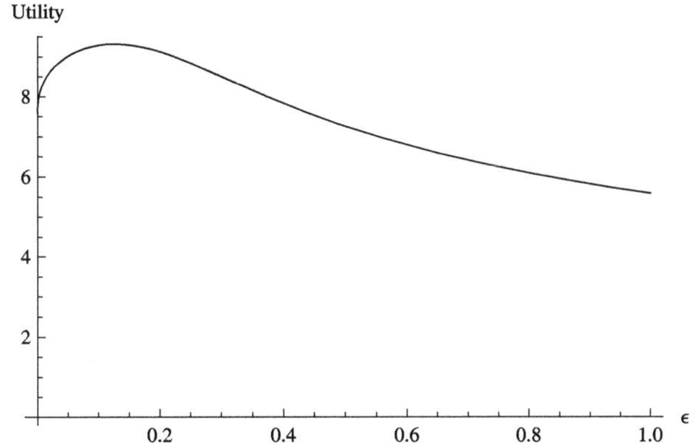

Fig. 15 Bifurcation diagram of equilibrium utility with respect to ε for $\alpha_1 = 0.6$, $\alpha_2 = 0.2$

convex-concave shape. At the beginning, the usage of G is quite low and does not change much with increasing ε. In this area, the abatement costs are still too low to change the advantage of conventional capital. The inflexion point is at $\varepsilon = 0.362$ where green capital starts to dominate conventional capital. From here on the ratio of G grows quite quickly until it converges to almost 100%. Note however, that 100% can never be reached. Accordingly, the ratio of K follows a concave-convex decrease.

In Fig. 17 the percentage values of the according R&D investments are depicted. Their development is similar, the only difference is the position of the inflexion point which is already at $\varepsilon = 0.263$.

5 Conclusion

The aim of this work is to investigate how environmental regulation influences economic growth as well as R&D investments and whether or not they induce a shift to a greener technology.

As far as economic growth is concerned, it becomes obvious that increasing stringency of environmental regulation causes a decline in both types of capital and

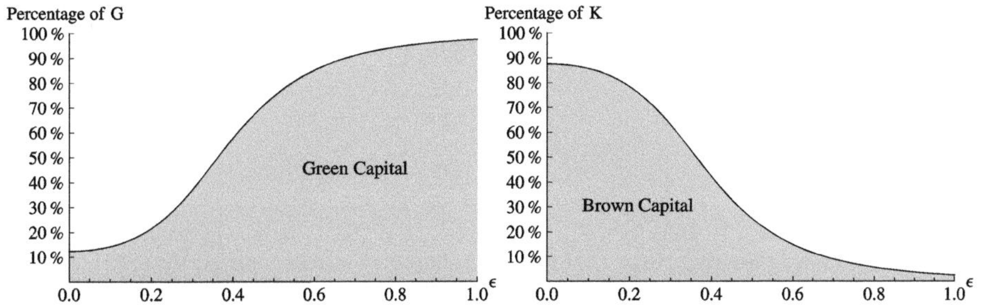

Fig. 16 Bifurcation diagram of the equilibrium percentage values of K and G with respect to ε for $\alpha_1 = 0.6$, $\alpha_2 = 0.2$

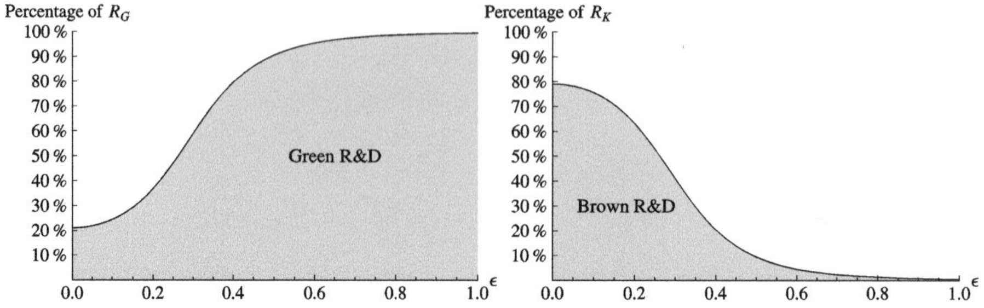

Fig. 17 Bifurcation diagram of the equilibrium percentage values of R_K and R_G with respect to ε for $\alpha_1 = 0.6$, $\alpha_2 = 0.2$

consequently also in production output. Therefore it rather represses than supports economic growth.

However, the analysis shows, that increasing environmental regulation indeed has a positive impact on the accumulation of green capital and on the increase of green R&D investments. This can especially be seen when the shares of capital levels and R&D investments under varying stringency of environmental standards are considered. Although both capital levels decline, increasing abatement costs even accelerate the decrease of brown capital levels so that in total production turns out to be greener the higher environmental quality standards are. The same applies for R&D investments.

To sum up, environmental regulation standards can cause a shift to greener production but only at the cost of reduced economic growth. Therefore, the introduction of additional environmental instruments, such as taxes or maybe subsidies, might be interesting and could possibly be helpful to achieve better results.

We want to close this paper with pointing out two further model extensions we would like to consider for future work. First, environmental quality so far is determined exogenously through the required standards set by the government. Hence, an interesting aspect would be to include an emission function $E(K, G)$ describing the pollution during the production process. Second, a main assumption in the present model is that the abatement effort exactly equals the necessary level needed to satisfy the required standards. This, however, shall be adapted for future model approaches by considering the abatement share as third control so that an environmentally aware agent can abate even more than necessary for the standards.

References

Acemoglu, D., Aghion, P., Bursztyn, L., & Hemous, D. (2009). *The environment and directed technical change* (Working Paper 15451). NBER. http://www.nber.org/papers/w15451.

Antweiler, W., Copeland, B. R., & Taylor, M. S. (2001). Is free trade good for the environment? *American Economic Review*, *91*(4), 877–908.

Bretschger, L., & Smulders, S. (2007). Sustainable resource use and economic dynamics. *Environmental & Resource Economics*, *36*(1), 1–13.

Cassou, S. P., & Hamilton, S. F. (2004). The transition from dirty to clean industries: optimal fiscal policy and the environmental Kuznets curve. *Journal of Environmental Economics and Management*, *48*(3), 1050–1077.

Cunha-e-Sá, M. A., Leitão, A., & Reis, A. B. (2010). *Innovation and environmental policy: clean vs. dirty technical change* (FEUNL Working Paper Series No. wp548). Faculdade de Economia, Universidade Nova de Lisboa.

Grass, D., Caulkins, J. P., Feichtinger, G., Tragler, G., & Behrens, D. A. (2008). *Optimal control of nonlinear processes—with applications in drugs, corruption and terror*. Heidelberg: Springer.

Hartley, P. R., Medlock, K. B., Temzelides, T., & Zhang, X. (2010). *Innovation, renewable energy, and macroeconomic growth* (Working Paper). Baker Institute for Public Policy.

IPCC (2007). *Climate change 2007*. Cambridge: Cambridge University Press.

Kalkuhl, M., Edenhofer, O., & Lessmann, K. (2011). *Learning or lock-in: optimal technology policies to support mitigation* (CESifo Working Paper Series No. 3422). CESifo Group Munich.

Knoll, C., & Zuba, D. M. (2004). *Dynamic models of the US cocaine epidemic: modeling initiation and demand and computing optimal controls*. Ph.D. thesis, TU Wien.

Lange, A., & Moslener, U. (2004). A bird in the hand is worth two in the bush? When do we prefer something certainly dirty to something perhaps clean? *Journal of Risk and Uncertainty*, *29*, 35–51. 10.1023/B:RISK.0000031444.37497.26.

Lehmijoki, U., & Palokangas, T. (2010). Trade, population growth, and the environment in developing countries. *Journal of Population Economics*, *23*(4), 1351–1370.

Rasmussen, T. N. (2001). CO2 abatement policy with learning-by-doing in renewable energy. *Resource and Energy Economics*, *23*(4), 297–325.

Rauscher, M. (2009). *Green R&D versus end-of-pipe emission abatement: a model of directed technical change* (Thuenen-Series of Applied Economic Theory, Working Paper 106). Universität Rastock.

Richardson, K., et al. (2009). *Synthesis report from climate change: global risks, challenges and decisions*. Copenhagen: University of Copenhagen.

Saltari, E., & Travaglini, G. (2011). The effects of environmental policies on the abatement investment decisions of a green firm. *Resource and Energy Economics*, *33*(3), 666–685.

Xepapadeas, A. (1992). Environmental policy, adjustment costs, and behavior of the firm. *Journal of Environmental Economics and Management*, *23*(3), 258–275.

Optimal Control of Growth and Climate Change—Exploration of Scenarios

Helmut Maurer, Johann Jakob Preuß, and Willi Semmler

1 Introduction

Nordhaus (Nordhaus and Boyer 2000; Nordhaus 2008) has developed a dynamical model linking economic growth with climate change. This model represents the core of the DICE climate model which is extensively calibrated in his book (Nordhaus 2008). This canonical model has by now become a work horse of the research on the economics of climate change. The model variants presented here focus only on the core dynamic equations of the canonical model of growth and climate change. Though we refer to the Nordhaus Dice model as a point of reference, we work with a lower dimensional system. We have fewer equations but a more realistic modeling of the temperature dynamics. This simpler model variant allows us to explore in a transparent way policy options and permits to suggest some directions of future research.

The model considered here builds on the dynamical model developed by Greiner et al. (2010), who discuss multiple equilibria and thresholds in a canonical optimal control problem with infinite horizon. In this paper, we study various extensions of the basic optimal control problem and compare the solutions for finite horizon and infinite horizon. We admit terminal constraints for the state variable, consider the impacts of constraints (such as CO_2 and temperature constraints) on abatement

H. Maurer (✉) · J.J. Preuß
Institut für Numerische und Angewandte Mathematik, Universität Muenster, Einsteinstr. 62, Münster 48161, Germany
e-mail: maurer@math.uni-muenster.de

J.J. Preuß
e-mail: johann.preuss@t-online.de

W. Semmler
Economics Department, The New School for Social Research, 5th Avenue, New York, NY 10003, USA
e-mail: semmlerw@newschool.edu

J. Crespo Cuaresma et al. (eds.), *Green Growth and Sustainable Development*, Dynamic Modeling and Econometrics in Economics and Finance 14,
DOI 10.1007/978-3-642-34354-4_6, © Springer-Verlag Berlin Heidelberg 2013

policies and consumption, and try to adjust the temperature by suitable penalties on the temperature. Such constraints allow to explore the implications for mitigation policies arising from the Kyoto treaty (CO_2 constraint) and the Copenhagen agreement (temperature constraint). Overall, we understand the exploration of our different scenarios as guidance to different policy options.

The paper is organized as follows. Section 2 introduces the dynamic model of growth and climate change that will be called the canonical model. In Sect. 3, we formulate the basic optimal control problem associated with the canonical model. We consider several extensions of the basic control problem incorporating terminal conditions, a penalty functional on the temperature as well as control and state constraints. Section 4 discusses the evaluation of the necessary optimality conditions (Pontryagin Maximum Principle) for the different optimal control problems in Sect. 3. In particular, the adjoint equations allow us to compute the stationary points (steady states) of the canonical system which determine the behavior of the infinite-horizon optimal solution. Finally, in Sect. 5 we present a number of case studies illustrating the various types of optimal control problems in Sect. 3. Optimal control and state trajectories of infinite-horizon control problems are computed by the routine opttrj (Kunkel and von dem Hagen 2000), whereas solutions of finite-horizon control problems with control and state constraints are obtained by discretization and nonlinear programming methods (Betts 2010; Büskens and Maurer 2000; Wächter and Biegler 2006).

2 Dynamic Model of Growth and Climate Change

Our model starts with a basic growth model which includes a simplified dynamics of the link between economic growth and the earth's climate. For details of the model the reader can be referred to the model description in Nordhaus (2008) and Greiner et al. (2010). For basic facts on climate change, as much as it is caused by economic activity, we refer the reader to the work by Keller et al. (2000, 2004). In the basic model the economy is represented by a decision making household. Its consumption is chosen optimally over time. Greiner et al. (2010) treat only the case of discounted utility which is maximized over an infinite time horizon. In this paper also the case of a finite horizon will be treated. In contrast to Nordhaus, and Greiner et al. the case of how damages affect the household's welfare will also be studied as well as the cases of state constraints, for example temperature and CO_2 constraints.

2.1 Capital

The dynamics of the per capita capital is described by the following differential equation:

$$\dot{K} = Y - C - A - (\delta + n)K, \quad K(0) = K_0, \tag{1}$$

where Y is the per capita production, K the per capita capital, A the per capita abatement measure and δ the depreciation of capital. The input of labor, L, grows at a rate n. The per capita production Y is defined by the production function

$$Y = BK^{\alpha} D(T), \tag{2}$$

where $\alpha \in (0, 1)$ is the capital share and B a positive constant. The function $D(T)$ denotes the inverse of the damage, that results from an increase of the temperature T above the pre-industrial temperature T_o, and has the form

$$D(T) = \left(a_1 (T - T_o)^2 + 1\right)^{-\psi} \tag{3}$$

with $a_1 > 0$ and $\psi > 0$. This is called the *damage function* and its effect can be characterized as follows: The greater the deviation of the current temperature T from the pre-industrial temperature T_o, the smaller the function value $D(T)$ and accordingly the smaller the value of the per capita production Y.

2.2 Emission and CO_2 Concentration

It is assumed that economic activity emits greenhouse gases, which depend on the capital that is used for production and which are here given in CO_2 equivalents. Thus they can be understood as a function of the per capita capital K, relative to the per capita abatement measure A. A larger capital goes along with higher emissions. Formally, this results in the expression

$$E = \left(a \frac{LK}{LA}\right)^{\gamma} = (aK/A)^{\gamma} \tag{4}$$

for the emission, where L is the labor input and $\gamma > 0$ and $a > 0$ are constants. The bigger a, the bigger the emission for given K and A and accordingly the worse the corresponding technology for the environment.

Emission causes an increase of the greenhouse gas (CO_2 concentration) in the atmosphere. It develops according to the differential equation

$$\dot{M} = \beta_1 E - \mu M, \quad M(0) = M_0. \tag{5}$$

Here, μ is the inverse of the atmospheric lifetime of CO_2 and β_1 highlights the fact that a certain part of the greenhouse gas emission is captured by the oceans and does not reach the atmosphere.

2.3 Temperature

To model the climate system of the earth, an *energy balance model* is used; cf. Roedel and Wagner (2011). Some parameters in the following equations have been improved by discussions with W. Roedel (2011). The change of the average surface temperature T is given by the equation

$$c_h \frac{\mathrm{d}T}{\mathrm{d}t} = S_E - H - F_N, \quad T(0) = T_0. \tag{6}$$

All magnitudes on the right side indicate annual averages, so each time step has to include exactly one year, hence $\Delta t = 365 \cdot 24 \cdot 60 \cdot 60$ s $= 31536000$ s is assumed. Because of that the differential equation changes to

$$\dot{T} \equiv \frac{\mathrm{d}T}{\mathrm{d}t} = \frac{\Delta t}{c_h}(S_E - H - F_N), \quad T(0) = T_0. \tag{7}$$

The earth's surface is greatly covered by oceans. Its heat capacity is given by the numerical value $c_h = 210652078$ J/(m^2 K), that follows from the identity $c_h = 0.7\rho_w c_w d$, where $\rho_w = 1027$ kg/m^3 is the density and $c_w = 4186$ J/(kg K) the specific heat capacity of the sea water and $d = 70$ m describes the depth of the oceanic top layer where a mixing and thus a heat transport takes place. The factor 0.7 represents the proportion of sea water in the total surface of the earth. The unit of $\frac{\Delta t}{c_h}$ is given by

$$\frac{\mathrm{s}}{\mathrm{J/(m^2\,K)}} = \mathrm{s\,m^2\,K/J} = \mathrm{m^2\,K/W},$$

from which it follows that $\frac{\Delta t}{c_h} \approx 0.149707$ m^2 K/W.

S_E is the supplied sun energy, H the non-radiative energy flux and $F_N = F_\uparrow - F_\downarrow$ the net flux of the terrestrial radiation. $F_\uparrow$ complies with the Stefan Boltzmann law, which has the form

$$F_\uparrow = \varepsilon\sigma T^4 \tag{8}$$

with the relative emissivity $\varepsilon = 0.95$ and the Stefan Boltzmann constant $\sigma = 5.67 \cdot 10^{-8}$ W/(m^2 K^4). Furthermore, the flux ratio is $F_\uparrow/F_\downarrow = 116/97$ and the difference is $S_E - H = (1 - \alpha_1(T))\frac{Q}{4}$ with the solar constant $Q = 1367$ W/m^2 and the planetary albedo α_1, which indicates how much energy is reflected back to space. The factor $\frac{1}{4}$ is the ratio between the cross-sectional area πr_{earth}^2 and the surface area $4\pi r_{\mathrm{earth}}^2$ of the earth, because it receives the sun's radiation flux only on a hemisphere. The share of non-reflected sun energy is given by the differentiable function

$$1 - \alpha_1(T) = k_1 \frac{2}{\pi} \arctan\left(\frac{\pi(T - 293)}{2}\right) + k_2, \tag{9}$$

in which $k_1 = 5.6 \cdot 10^{-3}$ and $k_2 = 0.1795$ should apply.

A high concentration of greenhouse gases affects the temperature through the so-called radiative forcing, which describes the change of incoming and outgoing energy in the atmosphere. For carbon dioxide (CO_2) we have

$$F = 5.35 \ln\left(\frac{M}{M_o}\right) \ [\mathrm{W/m^2}] \tag{10}$$

Table 1 Parameter values in the order of appearance in (12) and (13)

$\rho = 0.035$,	$n = 0.03$,	$L_0 = 1$,	$B = 1$,	$\alpha = 0.18$,
$a_1 = 0.025$,	$T_o = 288$,	$\psi = 0.025$,	$\delta = 0.075$,	$\beta_1 = 0.49$,
$a = 3.5 \cdot 10^{-4}$,	$\gamma = 1$,	$\mu = 0.1$,	$\Delta t = 31536000$,	$c_h = 210652078$,
$k_1 = 5.6 \cdot 10^{-3}$,	$k_2 = 0.1795$,	$Q = 1367$,	$\varepsilon = 0.95$,	$\sigma = 5.67 \cdot 10^{-8}$,
$M_o = 1$.				

with the pre-industrial CO_2 concentration M_o. In summary, we obtain the following differential equation for the average surface temperature T,

$$\dot{T} = \frac{\Delta t}{c_h}\left(\left(1 - \alpha_1(T)\right)\frac{Q}{4} - \frac{19}{116}\varepsilon\sigma T^4 + 5.35 \ln\left(\frac{M}{M_o}\right)\right), \quad T(0) = T_0, \tag{11}$$

where the unit on the right hand side is given by $\mathrm{m}^2\,\mathrm{K/W} \cdot \mathrm{W/m}^2 = \mathrm{K}$.

3 Optimal Control Problems

We present several versions of optimal control problems associated with the dynamics (1), (5) and (7) which is considered on a time interval $[0, t_f]$ with terminal time $0 < t_f \leq \infty$. The state variable is the vector $X = (K, M, T) \in \mathbb{R}^3$, the control variable is given by $u = (C, A) \in \mathbb{R}^2$. The *basic optimal control problem* is defined as follows: determine a (piecewise continuous) control function $u = (C, A) : [0, t_f] \to \mathbb{R}^2$ that *maximizes* the objective (cost functional),

$$\max J(X, u) = \int_0^{t_f} \mathrm{e}^{-(\rho - n)t} L_0 \ln C \mathrm{d}t, \tag{12}$$

subject to the differential equations (1), (5), (7),

$$\begin{aligned} \dot{K} &= BK^\alpha D(T) - C - A - (\delta + n)K, \\ \dot{M} &= \beta_1 (aK/A)^\gamma - \mu M, \\ \dot{T} &= \frac{\Delta t}{c_h}\left(\left(1 - \alpha_1(T)\right)\frac{Q}{4} - \frac{19}{116}\varepsilon\sigma T^4\right) + 5.35 \ln\left(\frac{M}{M_o}\right), \end{aligned} \tag{13}$$

with initial conditions

$$K(0) = K_0, \qquad M(0) = M_0, \quad T(0) = T_0. \tag{14}$$

Recall the damage function (3) and albedo function (9):

$$\begin{aligned} D(T) &= \left(a_1 (T - T_o)^2 + 1\right)^{-\psi}, \\ 1 - \alpha_1(T) &= k_1 \frac{2}{\pi} \arctan\left(\frac{\pi(T - 293)}{2}\right) + k_2. \end{aligned}$$

A complete list of parameters can be found in Table 1.

The problem (12)–(14) is called a *finite–horizon* optimal control problem, if the terminal time is *finite*, $0 < t_f < \infty$, otherwise for $t_f = \infty$ it is called an *infinite–horizon* control problem.

Now we present some variants and extensions of the basic control problem. A simplified version of the control problem arises, when the abatement control is kept constant,

$$A(t) \equiv A_c \quad \text{for } 0 \le t \le t_f. \tag{15}$$

Then the consumption C is the only control variable. We shall also study terminal constraints for the state variable given by

$$K(t_f) \ge K_f, \qquad M(t_f) \le M_f, \qquad T(t_f) \le T_f, \tag{16}$$

with appropriate values K_f, M_f, T_f. In particular, a positive value $K_f > 0$ will prevent the capital from approaching zero. It is also of interest to impose *control constraints* of the form

$$C_{\min} \le C(t) \le C_{\max}, \qquad A_{\min} \le A(t) \le A_{\max} \quad 0 \le t \le t_f, \tag{17}$$

with suitable bounds $C_{\min} < C_{\max}$ and $A_{\min} < A_{\max}$. Another variant of the control problem is obtained when the objective functional (12) is modified by subtracting a penalty term which measures the quadratic deviation of the temperature $T(t)$ from a desirable temperature T_c,

$$\max J_T(X,u) = J(X,u) - c \int_0^{t_f} \big(T(t) - T_c\big)^2 dt \quad (c > 0). \tag{18}$$

Here, the negative sign of the penalty appears in the modified functional, since the penalty term will be minimized. Note that the penalty term does not involve a discount factor. The penalty term in the extended functional can be viewed as a so-called *soft state constraint*. From a practical point of view, it is more important to consider explicit *state constraints* of the form

$$S\big(X(t)\big) = S\big(K(t), M(t), T(t)\big) \ge 0 \quad \forall 0 \le t_s \le t \le t_f, \tag{19}$$

where the function $S : \mathbb{R}^3 \to \mathbb{R}$ is assumed to be sufficiently often differentiable. The starting time t_s for the state constraint can be positive, $t_s > 0$, to account for the fact that the state constraint may not be feasible at the initial time but should be satisfied on a terminal interval $[t_s, t_f]$.

We briefly review some basic notions for state constraints and refer the reader to Hartl et al. (1995), Maurer (1979) for a thorough theoretical discussion. A *boundary arc* is a subinterval $[t_1, t_2] \subset [t_s, t_f]$ with $S(X(t)) = 0$ for $t_1 \le t \le t_2$. If the interval $[t_1, t_2]$ is maximal with this property, then t_1 is called the *entry-time* and t_2 is called the *exit-time* of the boundary arc; t_1 and t_2 are also called *junction times*. A *contact point* $t_c \in (t_s, t_f)$ is defined by the condition that there exists $\varepsilon > 0$ such that

$$S\big(X(t_c)\big) = 0, \qquad S\big(X(t)\big) > 0 \quad \text{for } t_c - \varepsilon \le t < t_c \text{ and } t_c < t \le t_c + \varepsilon.$$

The occurrence of boundary arcs and contact points is closely related to the notion of the *order* $q \in \mathbb{N}_+$ *of a state constraint*. The index $q \in \mathbb{N}_+$ is defined as the lowest order time derivative of $S(X(T))$ that contains the control variable explicitly (Hartl et al. 1995; Maurer 1979). Specifically, we consider the following state constraints for K, M and T, which should hold jointly or separately:

$$S\big(X(t)\big) = K(t) - K_{\min} \geq 0 \quad \forall t_s \leq t \leq t_f, \tag{20}$$
$$S\big(X(t)\big) = M_{\max} - M(t) \geq 0 \quad \forall t_s \leq t \leq t_f, \tag{21}$$
$$S\big(X(t)\big) = T_{\max} - T(t) \geq 0 \quad \forall t_s \leq t \leq t_f. \tag{22}$$

It is straightforward to show that the state constraint (20) for K has the order $q = 1$, the constraint (21) for M has the order $q = 2$, and the constraint (22) for T has the order $q = 3$. State constraints of order $q = 1$ usually exhibit only boundary arcs and no contact points, whereas state constraints of order $q = 2$ can have both boundary arcs and contact points. For $q = 3$, there are no boundary arcs with an analytic junction, i.e., every junction with a boundary arc exhibits some kind of chattering. Examples for boundary arcs and contact points and the phenomenon of a non-analytic junction with a boundary arc $T(t) = T_{\max}$ will be discussed in Sect. 5.

4 Maximum Principle: Necessary Optimality Conditions

The celebrated Pontryagin Maximum Principle (Pontryagin et al. 1964; Hestenes 1966; Sethi and Thompson 2000) furnishes the necessary optimality conditions for the *finite-horizon* control problem (12)–(16). Maximum Principles for state constrained optimal control problems were discussed in Maurer (1979) and Hartl et al. (1995). The Maximum Principle for *infinite-horizon* control problems is presented in Aseev and Kryazhimskiy (2004, 2007), Michel (1982) and Seierstadt and Sydsaeter (1987). For a modern theory of infinite–horizon control problems we refer to Lykina (2010); Lykina et al. (2008).

4.1 Basic Control Problem

4.1.1 Steady States for Constant Abatement $A(t) = A_c$

First, we consider the case of a constant abatement control (15) with $A(t) \equiv A_c = 1.21 \cdot 10^{-3}$ for $0 \leq t \leq t_f$. Here, the consumption C is the only control variable. The current–value Hamiltonian (Pontryagin function) (cf. Aseev and Kryazhimskiy 2007; Seierstadt and Sydsaeter 1987; Sethi and Thompson 2000) is given by

$$\begin{aligned} H(X, \lambda, C) = \ln C &+ \lambda_K\big(BK^\alpha D(T) - C - A_c - (\delta + n)K\big) \\ &+ \lambda_M\big(\beta_1 a^\gamma K^\gamma A_c^{-\gamma} - \mu M\big) \\ &+ \lambda_T \frac{\Delta t}{c_h}\left(\big(1 - \alpha_1(T)\big)\frac{Q}{4} - \frac{19}{116}\varepsilon\sigma T^4 + 5.35 \ln\left(\frac{M}{M_o}\right)\right), \end{aligned} \tag{23}$$

where $\lambda = (\lambda_K, \lambda_M, \lambda_T)$ is the vector of adjoint variables (shadow prices). The adjoint differential equations $\dot{\lambda} = (\rho - n)\lambda - H_X$ read explicitly:

$$\begin{aligned}\dot{\lambda}_K &= (\rho+\delta)\lambda_K - \lambda_K \alpha K^{\alpha-1} B D(T) - \lambda_M \beta_1 \gamma a^\gamma K^{\gamma-1} A^{-\gamma},\\ \dot{\lambda}_M &= (\rho - n)\lambda_M + \lambda_M \mu - \lambda_T \frac{\Delta t}{c_h} 5.35 \frac{1}{M},\\ \dot{\lambda}_T &= (\rho - n)\lambda_T - \lambda_K B K^\alpha D'(T) + \lambda_T \frac{\Delta t}{c_h}\left(\frac{Q}{4}\alpha_1'(T) + \frac{19}{116}\varepsilon\sigma 4T^3\right).\end{aligned} \tag{24}$$

The derivatives of the albedo function $\alpha_1(T)$ and the damage function $D(T)$ are

$$\begin{aligned}\alpha_1'(T) &= -5.6 \cdot 10^{-3}\left(1 + 0.25\pi^2 (T - 293)^2\right)^{-1}\\ D'(T) &= -2a_1\psi(T - T_o)\left(a_1(T - T_o)^2 + 1\right)^{-\psi-1}.\end{aligned} \tag{25}$$

The control C maximizes the Hamiltonian (23). Since no control constraints are imposed, we get the condition $H_C = 1/C - \lambda_K = 0$ implying

$$C = \frac{1}{\lambda_K} \quad \text{or} \quad \lambda_K = \frac{1}{C}. \tag{26}$$

Note that the *strict Legendre-Clebsch condition* is satisfied in view of

$$H_{CC} = -1/C^2 < 0.$$

The two expressions in (26) lead to two different systems of differential equations that contain either the control C or the adjoint variable λ_K. In this paper, we use the expression $C = 1/\lambda_K$ and work with the adjoint equations (24), whereas Greiner et al. (2010) choose $\lambda_K = 1/C$ to eliminate λ_K.

Thus with $C = 1/\lambda_K$, the state equations (13) and the adjoint equations (24) constitute a system of six differential equations. To calculate the steady states (stationary points) of this system, we consider the nonlinear equation of order six,

$$F(X,\lambda)^* = \left(\dot{X}^*, \dot{\lambda}\right) = 0 \in \mathbb{R}^6, \tag{27}$$

where * denotes the transpose. To solve this equation we proceed as follows

1. $\dot{\lambda}_M = 0$ is solved for $M = M(\lambda_M, \lambda_T, \cdot)$,
2. $\dot{M} = 0$ is solved for $\lambda_T = \lambda_T(K, T, \lambda_M, \cdot)$,
3. $\dot{\lambda}_K = 0$ is solved for $\lambda_K = \lambda_K(K, T, \lambda_M, \cdot)$ and finally
4. $\dot{K} = 0$ is solved for $\lambda_M = \lambda_M(K, T, \cdot)$.

In this way, we eliminate the variables M and λ in the equation (27) and are left with two equations for $\dot{T}$ and $\dot{\lambda}_K$ that depend only on the variables T and K. Figure 1(a) shows that the isoclines $\dot{T} = 0$ and $\dot{\lambda}_K = 0$ have three intersection points, each of them corresponding to a steady state. Numerical values of the three steady states are found in Table 2.

Stability properties of the three steady states are determined by the eigenvalues of the Jacobian of the function $F(X,\lambda)$ in (27) evaluated at the steady states. The

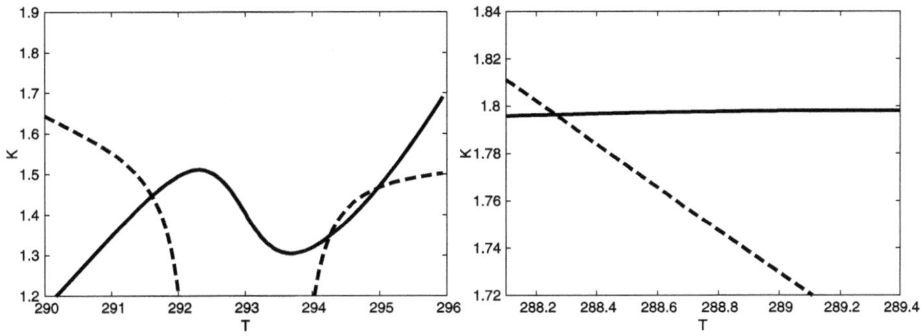

Fig. 1 Isoclines for $\dot{T} = 0$ (*solid*) and $\dot{\lambda}_T = 0$ (*dashed*). (*left*) constant abatement $A(t) \equiv 1.21 \cdot 10^{-3}$, (*right*) social optimum for control $u = (C, A)$

Table 2 Steady states for abatement $A(t) \equiv 1.21 \cdot 10^{-3}$

	Steady state I	Steady state II	Steady state III
K	1.4471998	1.3472580	1.4660558
M	2.0511964	1.9095434	2.0779221
T	291.60713	294.25816	294.96943
λ_K	1.1011613	1.1178274	1.1172386
λ_M	−0.17093840	−0.22542515	−0.14931998
λ_T	−0.045966518	−0.056432593	−0.040676338
C	0.90813214	0.89459250	0.89506393

Jacobian has six eigenvalues that are listed in Table 3. Since the real parts of the eigenvalues are nonzero, every steady state is *hyperbolic*. The first and third steady state have three eigenvalues with a positive and three eigenvalues with a negative real part, which implies that they are *saddle points*. However, the second steady state has only two eigenvalues with a negative real part, hence, it is unstable but has a two-dimensional stable manifold.

Table 3 Eigenvalues of the Jacobian of F, $A = 1.21 \cdot 10^{-3}$

Steady state I	Steady state II	Steady state III
−0.270660	0.283712	−0.260618
0.275660	−0.278712	0.265618
−0.094078±0.059941i	−0.136197	−0.114440
	0.141197	−0.067721
0.099078±0.059941i	0.002500±0.072057i	0.119440
		0.072721

4.2 Social Optimum for Control $u = (C, A)$

The current-value Hamiltonian for the optimal control problem with two control variables (C, A) agrees with that in (23) except that now the abatement A is a variable,

$$
\begin{aligned}
&H(X, \lambda, C, A) \\
&\quad = \ln C + \lambda_K \big(BK^\alpha D(T) - C - A - (\delta + n)K\big) \\
&\qquad + \lambda_M \big(\beta_1 a^\gamma K^\gamma A^{-\gamma} - \mu M\big) \\
&\qquad + \lambda_T \frac{\Delta t}{c_h}\left(\big(1 - \alpha_1(T)\big)\frac{Q}{4} + \frac{19}{116}\varepsilon\sigma T^4 + 5.35 \ln\left(\frac{M}{M_o}\right)\right).
\end{aligned}
\tag{28}
$$

The adjoint equations $\dot{\lambda} = (\rho - n)\lambda - H_X$ are identical with (23). The controls C and A that maximize the Hamiltonian are determined by the conditions

$$H_C = 1/C - \lambda_K = 0, \qquad H_A = -\gamma \lambda_M \beta_1 a^\gamma K^\gamma A^{-\gamma-1} - \lambda_K = 0,$$

which implies

$$C = \frac{1}{\lambda_K}, \qquad A = \left(-\gamma \frac{\lambda_M}{\lambda_K} \beta_1 a^\gamma K^\gamma\right)^{1/(1+\gamma)}. \tag{29}$$

The second derivatives of H are given by $H_{CA} = 0$ and

$$
\begin{aligned}
&H_{CC} = -\frac{1}{C^2} < 0, \\
&H_{AA} = \gamma(\gamma + 1)\lambda_M \beta_1 a^\gamma K^\gamma A^{-\gamma-2} < 0 \quad \text{for } \lambda_M < 0.
\end{aligned}
\tag{30}
$$

Note that the strict Legendre–Clebsch condition $H_{uu} < 0$ is only satisfied if $\lambda_M < 0$ holds. This sign condition will be verified in all examples in the next section. It follows from the control representation (29) that the optimal control $u = (C, A)$ is a *continuous* and even an analytic function.

The steady state calculation proceeds as above. Here, one substitutes the control terms (29) into the state equation (13) and adjoint equation (24), and thus obtains as in (27) a six-dimensional equation

$$F(X, \lambda)^* = \big(\dot{X}^*, \dot{\lambda}\big) = 0 \in \mathbb{R}^6.$$

In this case, one finds only a single steady state; see Figure 1(b) and Table 4. The six eigenvalues of the Jacobian of $F(X, \lambda)$ at the steady state are computed as

$$
\begin{aligned}
&-0.213640, \qquad 0.218640, \\
&-0.162069 \pm 0.129951i, \qquad 0.167069 \pm 0.129951i.
\end{aligned}
$$

There are three eigenvalues with a positive and three eigenvalues with a negative real part. Therefore, the steady state is a *saddle point*.

Table 4 Steady state for control (C, A): Social optimum

K	1.7964682	λ_K	1.0868071
M	1.2859989	λ_M	−0.020246732
T	288.28653	λ_T	−0.0034134235
C	0.92012653	A	0.0023957545

4.3 Transversality Conditions for Adjoint Variables

In the basic control problem, no terminal state conditions were prescribed. In the *finite-horizon* case, the transversality for the adjoint variables is

$$\lambda(t_f) = \big(\lambda_K(t_f), \lambda_M(t_f), \lambda_T(t_f)\big) = (0, 0, 0).$$

Note that the condition $\lambda_K(t_f) = 0$ is incompatible with the control law $C(t) = 1/\lambda_K(t)$. As consequence, in order to get a well-defined solution one has to impose either a terminal constraint $K(t_f) \geq K_f > 0$ or a control constraint $C(t) \leq C_{\max}$; cf. Sect. 5.5.

This is not relevant when studying *infinite-horizon* optimal control problems. Here, the adjoint variable $\lambda(t)$ converges to one of the steady states. The transversality condition at infinity then takes the form (Aseev and Kryazhimskiy 2004; Aseev and Kryazhimskiy 2007; Michel 1982; Sethi and Thompson 2000),

$$\lim_{t\to\infty} \mathrm{e}^{-(\rho-n)t}\lambda(t) = 0. \tag{31}$$

When the terminal constraints (16)

$$K(t_f) \geq K_f, \qquad M(t_f) \leq M_f, \qquad T(t_f) \leq T_f,$$

are imposed in the *finite-horizon* control problem, the transversality condition for adjoint variables asserts that there exist multipliers $\nu_K, \nu_M, \nu_T \in \mathbb{R}$ with

$$\begin{aligned} \lambda_K(t_f) &= \nu_K \geq 0, \quad \nu_K\big(K(t_f) - K_f\big) = 0, \\ \lambda_M(t_f) &= \nu_M \leq 0, \quad \nu_M\big(M(t_f) - M_f\big) = 0, \\ \lambda_T(t_f) &= \nu_T \leq 0, \quad \nu_T\big(T(t_f) - T_f\big) = 0. \end{aligned} \tag{32}$$

Recall that in the *infinite-horizon* case we can not prescribe terminal conditions, since the trajectory converges to one of the steady states.

4.4 Control Constraints

In the case of the control constraints (17),

$$C_{\min} \leq C(t) \leq C_{\max}, \qquad A_{\min} \leq A(t) \leq A_{\max} \quad \forall t \in [0, t_f],$$

the control expressions (17) have to be replaced by the projections onto the control sets,

$$C = \mathrm{proj}_{[C_{\min}, C_{\max}]}(1/\lambda_K), \qquad A = \mathrm{proj}_{[A_{\min}, A_{\max}]}\left(-\gamma\frac{\lambda_M}{\lambda_K}\beta_1 a^\gamma\right). \tag{33}$$

4.5 State Constraints

In (19), we considered the general state constraint

$$S\big(X(t)\big) = S\big(K(t), M(T), T(t)\big) \geq 0 \quad \forall 0 \leq t_s' t \leq t_f.$$

Practically relevant state constraints were considered in (20)–(22),

$$\begin{aligned} S\big(X(t)\big) &= K(t) - K_{\min} \geq 0 \quad \forall t_s \leq t \leq t_f, \\ S\big(X(t)\big) &= M_{\max} - M(t) \geq 0 \quad \forall t_s \leq t \leq t_f, \\ S\big(X(t)\big) &= T_{\max} - T(t) \geq 0 \quad \forall t_s \leq t \leq t_f. \end{aligned} \tag{34}$$

To evaluate necessary optimality conditions, we use the *direct adjoining approach* described in Maurer (1979), Hartl et al. (1995), where the state constraint is directly adjoined to the Hamiltonian by a multiplier μ which defines the *augmented Hamiltonian*

$$\mathcal{H}(X, \lambda, \mu, C, A) = H(X, \lambda, C, A) + \mu S(X)$$

Under some additional regularity conditions, the Maximum Principle (Maurer 1979; Hartl et al. 1995) asserts that there exists a multiplier function $\mu : [0, t_f] \to \mathbb{R}_+$ such that the adjoint variables λ satisfies the adjoint equation

$$\dot{\lambda} = (\rho - n)\lambda - \mathcal{H}_X = (\rho - n)\lambda - H_X - \mu S_X \tag{35}$$

and the complementarity condition $\mu(t)S(X(t)) = 0 \ \forall t \in [0, t_f]$ holds. Moreover, at every contact or junction point t_1, the adjoint variable may have a jump according to

$$\lambda\big(t_1^+\big) = \lambda\big(t_1^-\big) - \nu_1 S_X\big(X(t_1)\big), \quad \nu_1 \geq 0. \tag{36}$$

For the state constraints (34), we get the jump conditions

$$\begin{aligned} \lambda_K\big(t_1^+\big) &= \lambda_K\big(t_1^-\big) - \nu_K, \quad \nu_K \geq 0, \\ \lambda_M\big(t_1^+\big) &= \lambda_M\big(t_1^-\big) - \nu_M, \quad \nu_M \geq 0, \\ \lambda_T\big(t_1^+\big) &= \lambda_T\big(t_1^-\big) - \nu_T, \quad \nu_T \geq 0. \end{aligned} \tag{37}$$

5 Numerical Results for Various Scenarios

5.1 Numerical Methods

We have used direct optimization methods for solving the *finite-horizon* basic optimal control problem (12)–(14) and its extension incorporating the constraints or a modified functional (16)–(22). The direct optimization approach is based on a suitable discretization of the control problem by which the control problem is transcribed into a (large-scale) nonlinear programming problem (NLP). Such NLP can efficiently be solved either by Sequential Quadratic Programming (SQP) methods (cf. Betts 2010 and Büskens and Maurer 2000) or by an Interior-Point method like

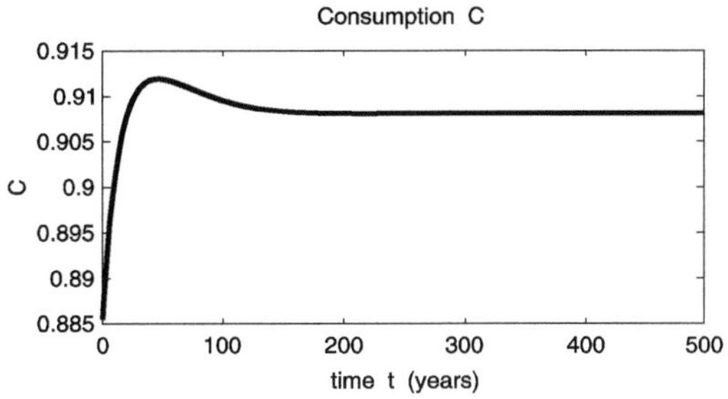

Fig. 2 *Infinite horizon, abatement* $A_c = 1.21 \cdot 10^{-3}$, $T(0) = 290$. Consumption C

IPOPT (cf. Wächter and Biegler 2006). It is very convenient to formulate the discretized control problem by means of the modeling language AMPL developed by Fourer et al. (1993). It can be shown that the Lagrange multipliers of the NLP represent the adjoint variables $\tilde{\lambda}(t)$ for the discounted objective (12). Then the adjoint variables in the current-value formulation are obtained as $\lambda(t) = \exp((\rho - n)t)\tilde{\lambda}(t)$.

To solve the *infinite-horizon* optimal control problem we implemented the solver OPTTRJ developed by Kunkel and von dem Hagen (2000). In this approach, a boundary value problem for the state and adjoint variable $(X, \lambda) \in \mathbb{R}^6$ is solved, where the dynamic equations are given by (13) and (23) and the control variables are substituted by the expressions (26) or (29). By a suitable time transformation, the infinite time interval $[0, \infty)$ is transformed into the finite time interval $[0, 1]$. Terminal conditions for state and adjoint variables are determined by the eigenvalues of the Jacobian of the mapping $F(X, \lambda)$ in (27) evaluated at the steady states.

We shall start the numerical analysis with the *infinite-horizon* control problem, since their optimal trajectories serve as a point of orientation for the *finite-horizon* case. We shall see that the *finite-horizon* case offers a greater flexibility in handling terminal constraints and control or state constraints.

5.2 *Infinite Horizon, Abatement* $A_c = 1.21 \cdot 10^{-3}$ *and* $T(0) = 290$

For the initial condition

$$T(0) = 290, \qquad K(0) = 1.4, \qquad M(0) = 2.0,$$

the infinite horizon solution converges to the steady state I in Table 2. The control and state and adjoint variables are shown in Figs. 2 and 3 on the time interval $[0, 500]$. The code OPTTRJ Kunkel and von dem Hagen (2000) yields the following numerical results

$$\begin{aligned} X(\infty) &= (1.4471997, 2.0511964, 291.60713), \\ \lambda(0) &= (1.1293319, -0.13696052, -0.030260502), \\ \lambda(\infty) &= (1.1011613, -0.17093840, -0.045966518). \end{aligned}$$

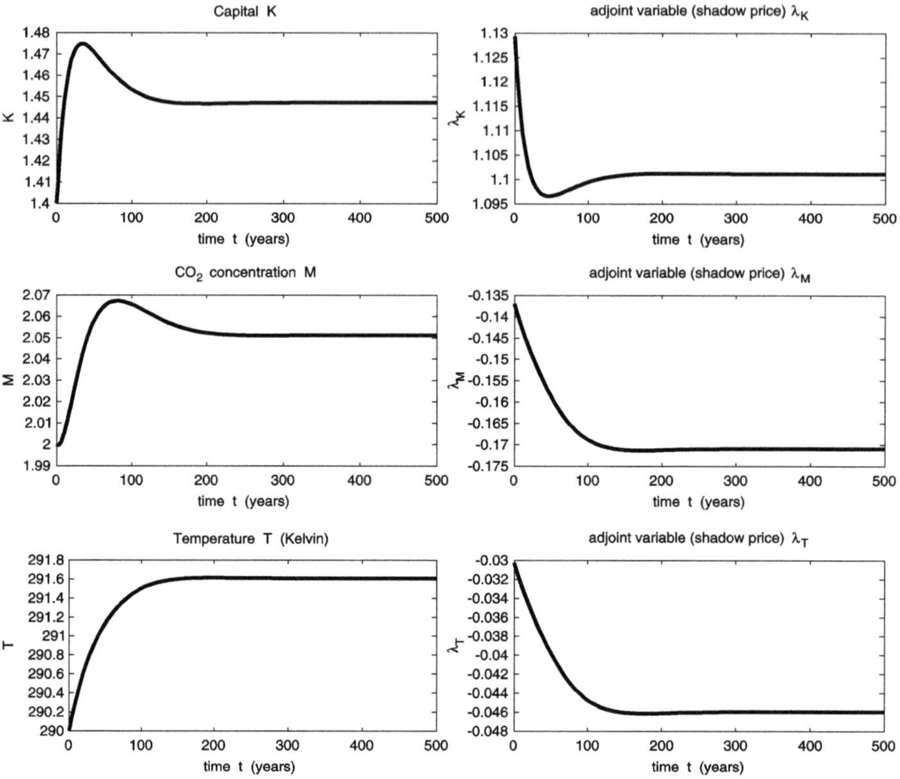

Fig. 3 *Infinite horizon, abatement* $A_c = 1.21 \cdot 10^{-3}$, $T(0) = 290$. *Top row*: capital K and adjoint variable λ_K. *Middle row*: CO_2 concentration M and adjoint variable λ_M. *Bottom row*: temperature T and adjoint variable λ_T

5.3 Infinite Horizon, Abatement $A_c = 1.21 \cdot 10^{-3}$ and $T(0) = 293$

We chose the initial condition

$$T(0) = 293, \qquad K(0) = 1.4, \qquad M(0) = 2.0$$

with a rather high initial temperature. Even in this case, the infinite horizon solution converges to the steady state I in Table 2. The control and state variables are shown in Fig. 4 on the time interval [0, 500]. The code OPTTRJ (Kunkel and von dem Hagen 2000) gives the numerical results

$$\begin{aligned} X(\infty) &= (1.4471997, 2.0511964, 291.60713), \\ \lambda(0) &= (0.98130594, -0.67165995, -0.041997270), \\ \lambda(\infty) &= (1.1011613, -0.17093840, -0.045966518). \end{aligned}$$

It is noteworthy that even for the higher initial temperature $T(0) = 294$ the optimal trajectories converge to the steady state I and are similar to those in Fig. 4.

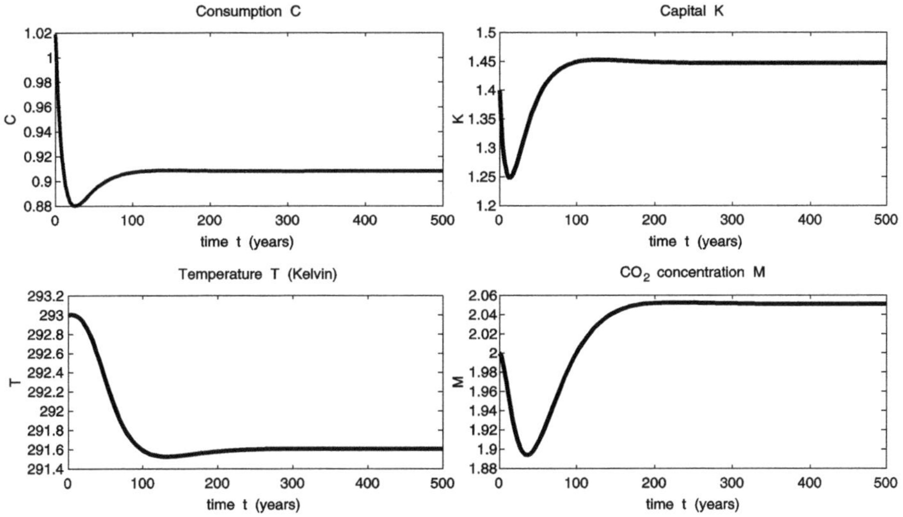

Fig. 4 *Infinite horizon, abatement* $A_c = 1.21 \cdot 10^{-3}$, $T(0) = 293$. *Top row*: consumption C and capital K. *Bottom row*: temperature T and CO_2 concentration M

Thus, despite high initial temperatures there exist infinite-horizon solutions that are not doomed to converge to the steady state III in Table 2 with the high final temperature $T = 294.969$.

5.4 Infinite Horizon: Social Optimum for Control $u = (C, A)$

Again, we consider the initial condition

$$T(0) = 292, \qquad K(0) = 1.4, \qquad M(0) = 2.0.$$

Using both controls C and A the infinite horizon solution converges to the steady state in Table 3 and thus terminates slightly above the pre-industrial temperature $T_o = 288$.

The controls C and A are shown in Fig. 5, while the state and adjoint variables are depicted in Fig. 6. We obtain the numerical results

$$X(\infty) = (1.7964682, 1.2859989, 288.28652),$$
$$\lambda(0) = (1.1144233, -0.051097441, -0.017504689),$$
$$\lambda(\infty) = (1.0868071, -0.020246782, -0.0034134235).$$

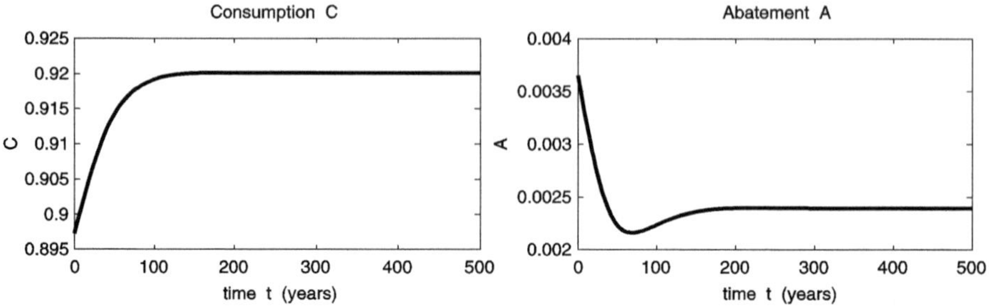

Fig. 5 *Infinite horizon, social optimum*, $T(0) = 292$. Consumption C and abatement A

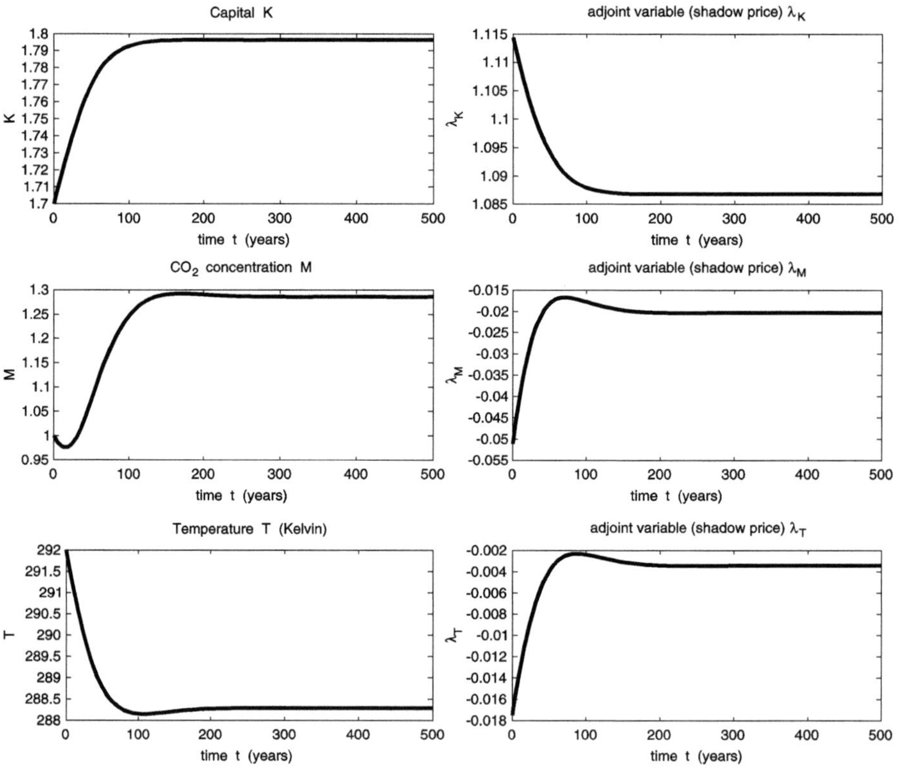

Fig. 6 *Infinite horizon, social optimum*, $T(0) = 292$. *Top row*: capital K and adjoint variable λ_K. *Middle row*: CO_2 concentration M and adjoint variable λ_M. *Bottom row*: temperature T and adjoint variable λ_T

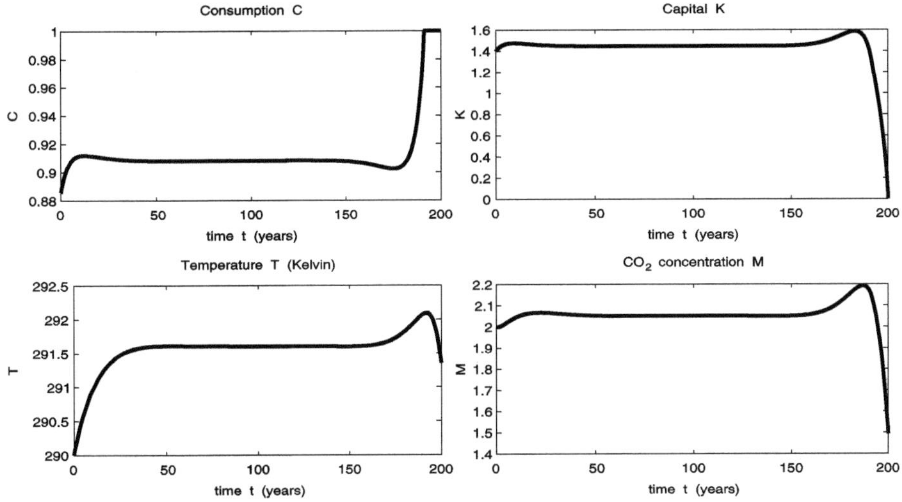

Fig. 7 *Finite horizon* $t_f = 200$, *abatement* $A_c = 1.21 \cdot 10^{-3}$, $T(0) = 290$. *Top row*: consumption C and capital K. *Bottom row*: temperature T and CO_2 concentration M

5.5 Finite Horizon: Basic Control Problem with Abatement $A_c = 1.21 \cdot 10^{-3}$

The initial condition are

$$T(0) = 290, \qquad K(0) = 1.4, \qquad M(0) = 2.0.$$

Since no terminal conditions are prescribed, a control constraint has to be imposed. Otherwise the control law $C = 1/\lambda_K$ can not be applied due to $\lambda_K(t_f) = 0$. We choose the control constraint

$$C(t) \le 1 \quad \forall 0 \le t \le t_f.$$

The code IPOPT provides the control C and state variables displayed in Fig. 7 and the numerical results

$$\begin{aligned} J(X,u) &= -11.7972, \\ X(t_f) &= (0.000445590, 1.49287, 291.354), \\ \lambda(0) &= (1.12933, -0.136961, -0.030260), \\ \lambda(t_f) &= (0.174102, 0.0, 0.0). \end{aligned}$$

Fig. 8 *Finite horizon* $t_f = 200$: *abatement* $A_c = 1.21 \cdot 10^{-3}$, $T(0) = 292$, *terminal constraint* $X(t_f) = X_{s1}$. Consumption C

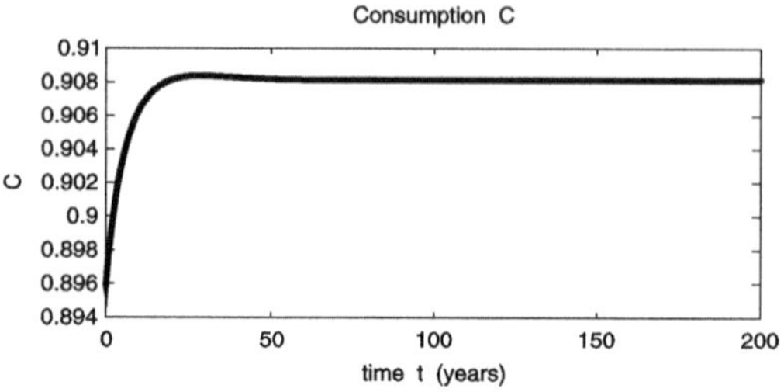

5.6 Finite Horizon, Abatement $A_c = 1.21 \cdot 10^{-3}$ and Terminal Condition $X(t_f) = X_{s,1}$

To avoid the strong decrease of capital and increase of consumption in Fig. 7 the basic control problem, we prescribe the steady state I in Table 2 as a terminal condition and choose the boundary conditions

$$T(0) = 292, \qquad K(0) = 1.4, \qquad M(0) = 2,$$
$$X(t_f) = X_{s1} = (1.4471998, 2.0511964, 291.60713)$$

The solution is displayed in Figs. 8 and 9. The code IPOPT gives the results

$$J(X,u) = -12.2455$$
$$\lambda(0) = (1.11619, -0.180347, -0.0522408)$$
$$\lambda(t_f) = (1.10116, -0.170933, -0.0459687)$$

5.7 Finite Horizon: Abatement $A_c = 1.21 \cdot 10^{-1}$, $T(0) = 292$, $T(t_f) = 290$

It is desirable to reach a smaller terminal temperature than the steady state temperature $T(t_f) = 291.607$ in the preceding case and attain a smaller CO_2 concentration M. Here, we choose the boundary conditions

$$T(0) = 292, \qquad K(0) = 1.4, \qquad M(0) = 2,$$
$$T(t_f) = 290, \qquad K(t_f) = 1.4, \qquad M(t_f) = 1.8.$$

The optimal trajectories computed by IPOPT are shown in Fig. 10. Numerical results of the functional value and the adjoint variables are

$$J(X,u) = -12.3439,$$
$$X(t_f) = (1.4, 1.683188, 290.0),$$
$$\lambda(0) = (1.11619, -0.180347, -0.0524082),$$
$$\lambda(t_f) = (1.13861, 0.0, -0.316547).$$

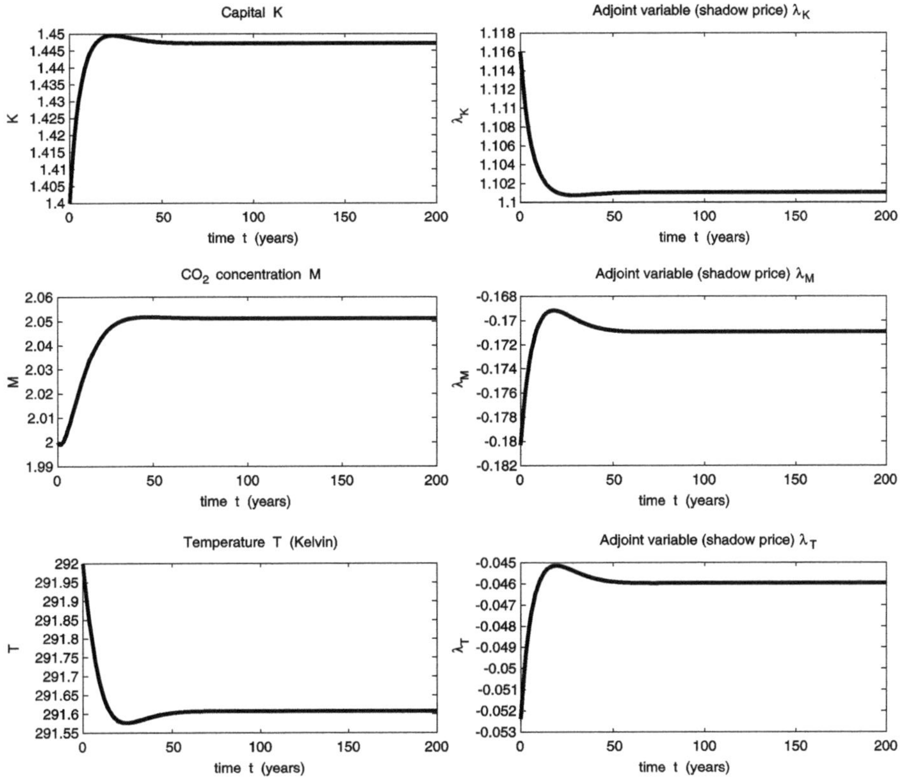

Fig. 9 *Finite horizon* $t_f = 200$, *abatement* $A_c = 1.21 \cdot 10^{-3}$, $T(0) = 292$, *terminal constraint* $X(t_f) = X_{s1}$. *Top row*: capital K and adjoint variable λ_K. *Middle row*: CO_2 concentration M and adjoint variable λ_M. *Bottom row*: temperature T and adjoint variable λ_T

The solution shows a strong decrease in capital and consumption. This effect can be avoided by imposing suitable control and state constraints; cf. the following scenario.

5.8 Finite Horizon: Abatement $A_c = 1.21 \cdot 10^{-1}$ and Control and State Constraints

This scenario treats the boundary conditions

$$T(0) = 292, \qquad K(0) = 1.4, \qquad M(0) = 2; \qquad T(t_f) = 290, \qquad K(t_f) = 1.3.$$

Motivated by Fig. 10, we impose control and state constraints,

$$0.895 \le C(t) \le 0.95, \qquad K(t) \ge 1.1, \qquad M(t) \le 1.8, \qquad t_s = 10 \le t \le t_f,$$

for which we obtain the numerical results

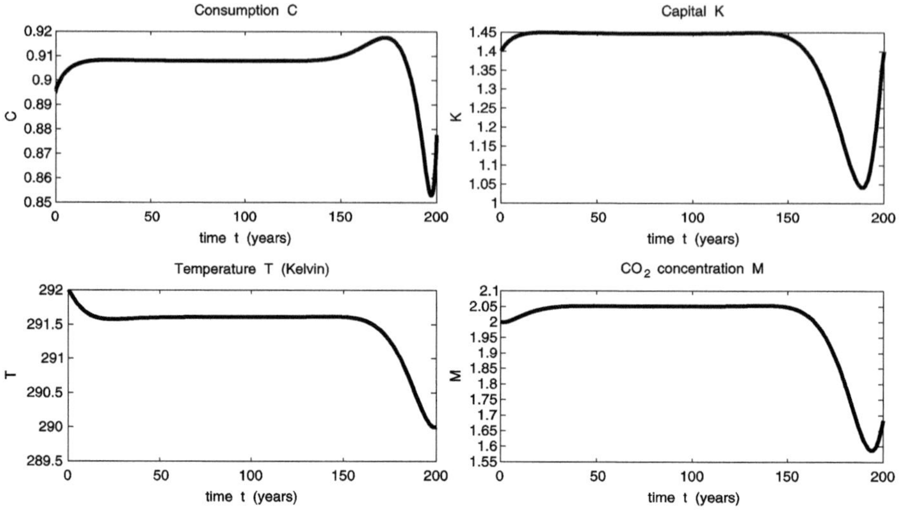

Fig. 10 *Finite horizon* $t_f = 200$, *abatement* $A_c = 1.21 \cdot 10^{-3}$, $T(0) = 292$, *terminal temperature* $T(t_f) = 290$. *Top row*: consumption C and capital K. *Bottom row*: temperature T and CO_2 concentration M

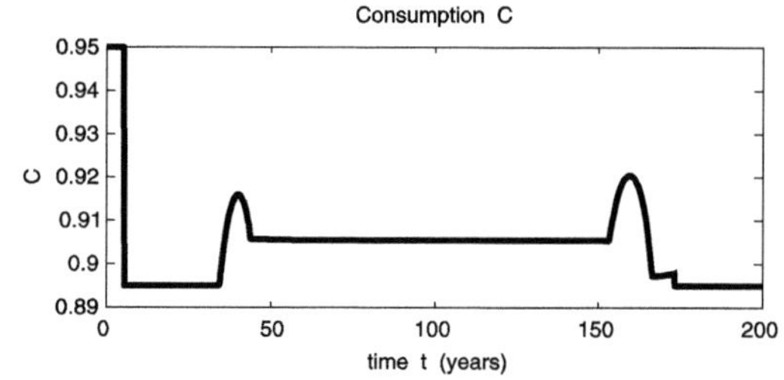

Fig. 11 *Finite horizon* $t_f = 200$: *abatement* $A_c = 1.21 \cdot 10^{-3}$, $T(0) = 292$, $T(t_f) = 290$, *constraints* $M(t) \leq 1.8$, $K(t) \geq 1.1$ and $0.895 \leq C(t) \leq 0.95$ for $t \geq 10$. Consumption C

$$
\begin{aligned}
J(X,u) &= -12.6479,\\
X(t_f) &= (1.3,\ 1.720185,\ 290.0),\\
\lambda(0) &= (0.0123934,\ -1.20090,\ -0.0452723),\\
\lambda(t_f) &= (3.17599,\ 0.0,\ -1.04181).
\end{aligned}
$$

The consumption C displayed in Fig. 11 has three boundary arcs, where the constraints $0.895 \leq C(t) \leq 0.95$ become active. The constraint $K(t) \geq 1.1$ is binding toward the end of the planning period. The associated adjoint variable λ_K is continuous though jumps are permitted according to the jump condition (37). This is due to the fact that this state constraint is of order $q = 1$, cf. Hartl et al. (1995). The state constraint $M(t) \leq 1.8$, $t \geq 10$, of order $q = 2$ becomes active at $t = t_s = 10$ and has a boundary arc in an intermediate interval $[t_1, t_2]$. Note that the adjoint variable $\lambda_K(t)$ has jumps at t_s and t_1, t_2. The optimal trajectories computed by IPOPT are shown in Fig. 12.

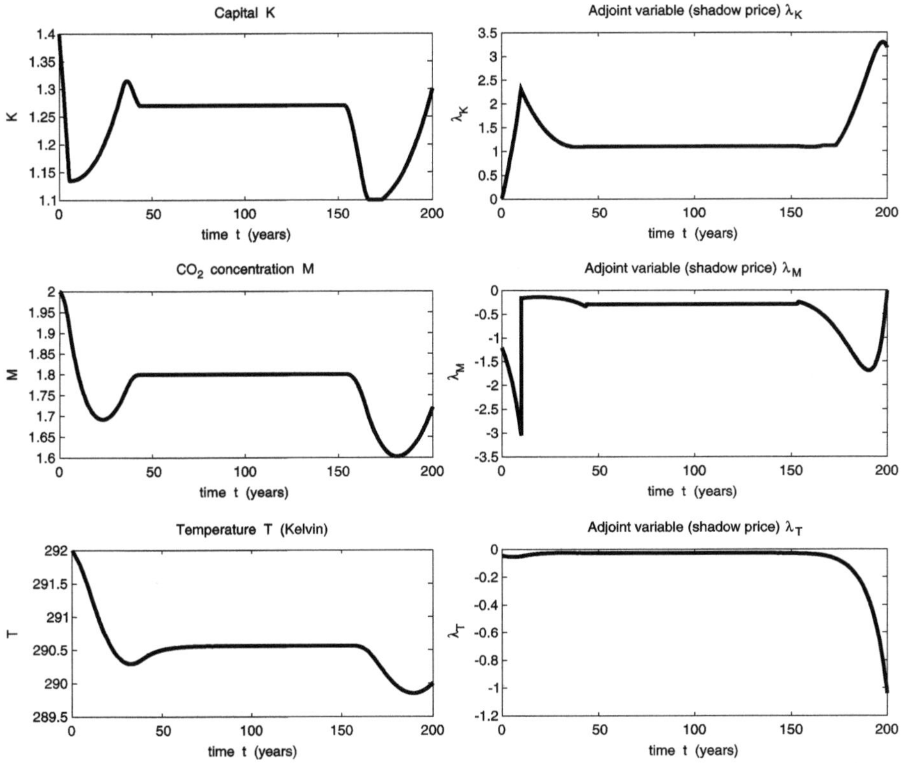

Fig. 12 *Finite horizon* $t_f = 200$, *abatement* $A_c = 1.21 \cdot 10^{-3}$, $T(0) = 292$, $T(t_f) = 290$, *constraints* $M(t) \leq 1.8$, $K(t) \geq 1.1$ and $0.895 \leq C(t) \leq 0.95$ for $t \geq 10$. *Top row*: capital K and adjoint variable λ_K. *Middle row*: CO_2 concentration M and adjoint variable λ_M. *Bottom row*: temperature T and adjoint variable λ_T

5.9 Finite Horizon: Abatement $A_c = 1.21 \cdot 10^{-1}$ *and State Constraint for T*

One may also efficiently decrease the initial temperature $T(0) = 292$ by imposing the state constraint

$$T(t) \leq 289 \quad \text{for } t_s = 10 \leq t \leq t_f.$$

With $K(0) = 1.4$, $M(0) = 2.0$ and the terminal constraint $K(t_f) = 1.3$ we get the numerical results

$$\begin{aligned}
J(X,u) &= -14.36051, \\
X(t_f) &= (1.3, 1.51783, 289.0), \\
\lambda(0) &= (0.847434, -0.755547, -0.0991925), \\
\lambda(t_f) &= (1.13135, 0.0, -0.358985).
\end{aligned}$$

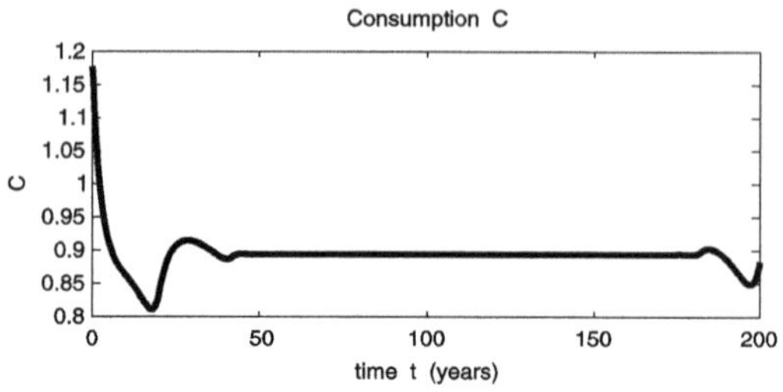

Fig. 13 *Finite horizon* $t_f = 200$, *abatement* $A_c = 1.21 \cdot 10^{-3}$ *and state constraint* $T(t) \leq 289$ *for* $t \geq t_s = 10$. Consumption C

The solution is displayed in Fig. 13. Figure 14 shows that the state constraint for T becomes active at $t = t_s = 10$ and on a boundary arc $[t_1, t_2]$. The adjoint variable $\lambda_K(t)$ has jumps at $t = 10$, t_1, t_2 in agreement with the jump condition (37). Since the state constraint has order $q = 3$, the junctions to the boundary arc are non-analytic which, however, can hardly be detected from the numerical solution.

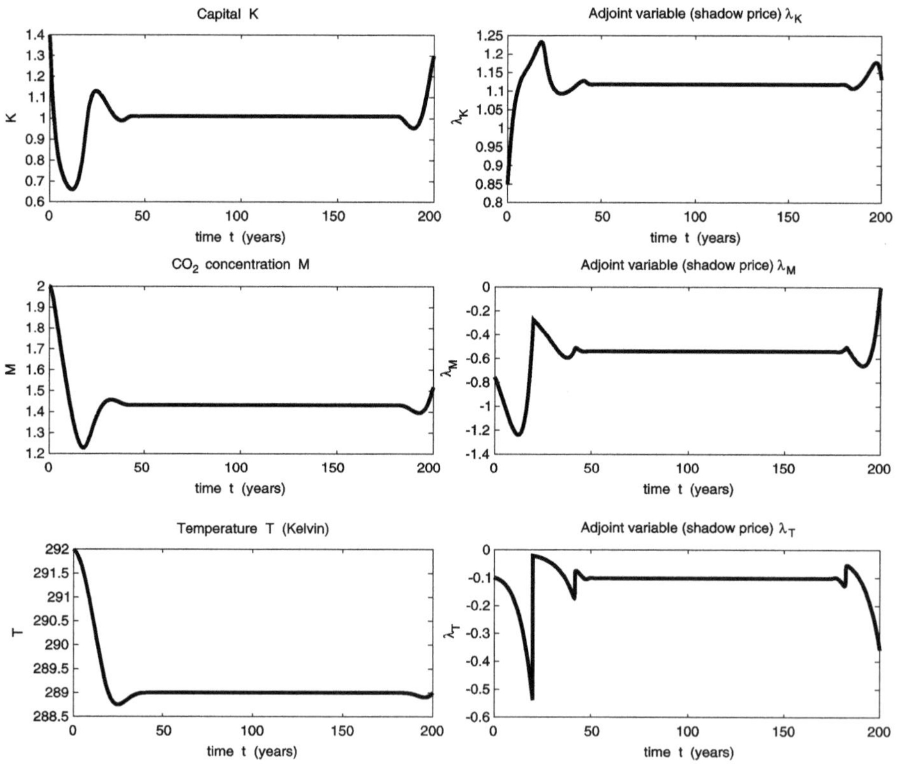

Fig. 14 *Finite horizon* $t_f = 200$, *abatement* $A_c = 1.21 \cdot 10^{-3}$ *and state constraint* $T(t) \leq 289$ $\forall t \geq 10$. *Top row*: capital K and adjoint variable λ_K. *Middle row*: CO_2 concentration M adjoint variable λ_M. *Bottom row*: temperature T and adjoint variable λ_T

5.10 Finite Horizon: Abatement $A_c = 1.21 \cdot 10^{-1}$ and Penalty Functional (18)

We make an attempt for adjusting the temperature during the control process by maximizing the penalty functional (18):

$$\max J_T(X,u) = \int_0^{t_f} e^{-(\rho-n)t} \ln C \mathrm{d}t - c_T \int_0^{t_f} \big(T(t) - T_c\big)^2 \mathrm{d}t \quad (c_T > 0).$$

We have to impose a lower bound for the capital; otherwise the capital tends to zero. For convenience, we also consider an upper bound for the consumption and thus impose the constraints

$$C(t) \le 1, \qquad K(t) \ge 1 \quad \text{for all } t \in [0, t_f].$$

We choose the initial temperature $T(0) = 292$ and try to get near the desired temperature $T_c = 289$ by choosing suitable penalty parameters c_T. In Fig. 15, the solutions for the penalty parameters $c_T = 0.01$ (left column) and $c_T = 0.001$ (right column) are compared. The left column in Fig. 15 shows that the aim of reaching the desired temperature $T_c = 289$ is quite well attained but goes at the expense of a decreasing consumption and capital level. A larger penalty does not significantly improve on this result, since the state constraint $K(t) \ge 1$ is an obstacle to further improvement.

The values of the cost functionals are

$$c_T = 0.01: \quad J(X,u) = -13.5105, \qquad J_T(X,u) = -14.7162,$$
$$c_T = 0.001: \quad J(X,u) = -12.9543, \qquad J_T(X,u) = -12.3949.$$

5.11 Finite Horizon: Social Optimum with Control $u = (C, A)$

Finally, we study the case of a social optimum using both controls $u = (C, A)$. We prescribe the steady state in Table 4 as terminal state. Hence, we choose the initial and terminal conditions

$$T(0) = 292, \qquad K(0) = 1.4, \qquad M(0) = 2.0;$$
$$X(t_f) = (1.796468, 1.285998, 288.2865).$$

Moreover, the following upper bound is imposed on the abatement control:

$$A(t) \le 0.003, \quad \forall 0 \le t \le t_f.$$

We obtain the following numerical results:

$$J(X,u) = -11.0766,$$
$$\lambda(0) = (1.20362, -0.0856009, -0.0330846),$$
$$\lambda(t_f) = (-1.08681, 0.0202467, -0.00341343).$$

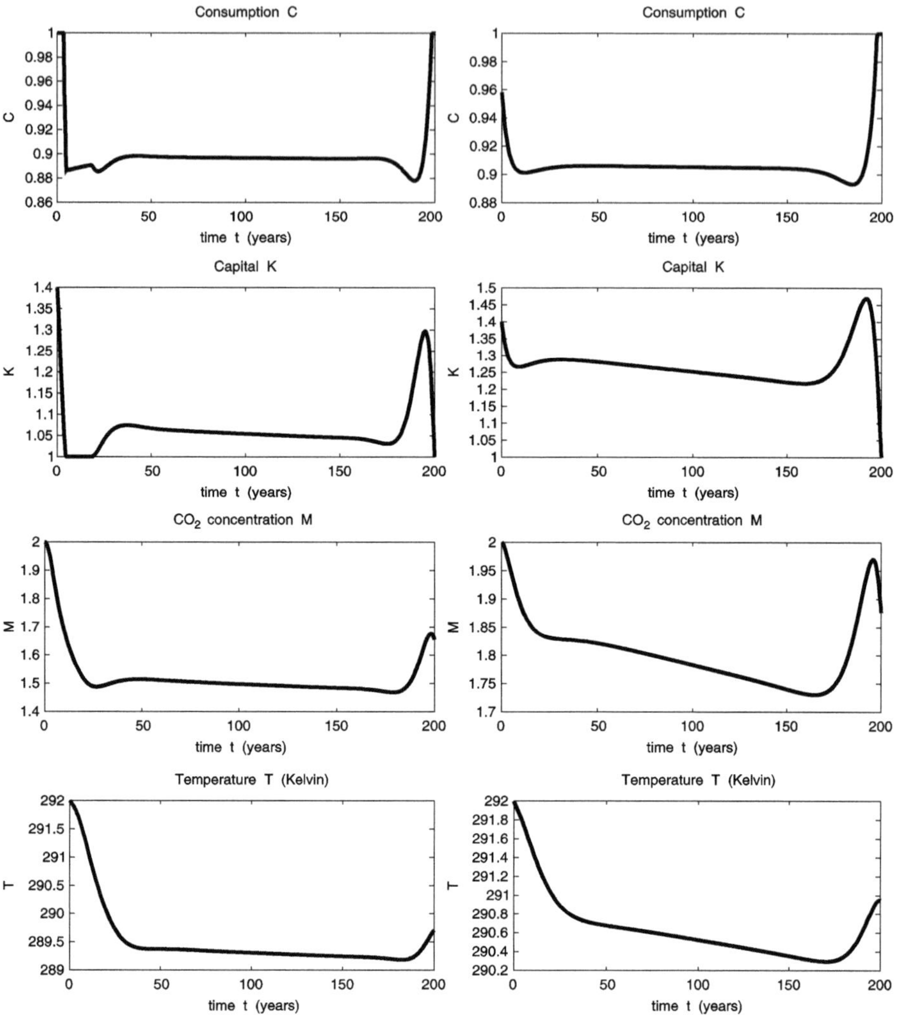

Fig. 15 *Finite horizon* $t_f = 200$, *abatement* $A_c = 1.21 \cdot 10^{-3}$, $T(0) = 292$ *and penalty (18). Left column*: penalty $c_T = 0.01$. *Right column*: penalty $c_T = 0.001$. Consumption C, capital K, CO_2 concentration M and temperature T

Figure 16 displays the control and state variables for the initial temperature $T(0) = 292$; it clearly reflects the fact that the maximum abatement is needed for at least 13 years to substantially decrease the temperature T and CO_2 concentration M. However, it is remarkable that the decrease in temperature is more pronounced in the finite-horizon solution than in the infinite-horizon solution displayed in Fig. 16, bottom row (b).

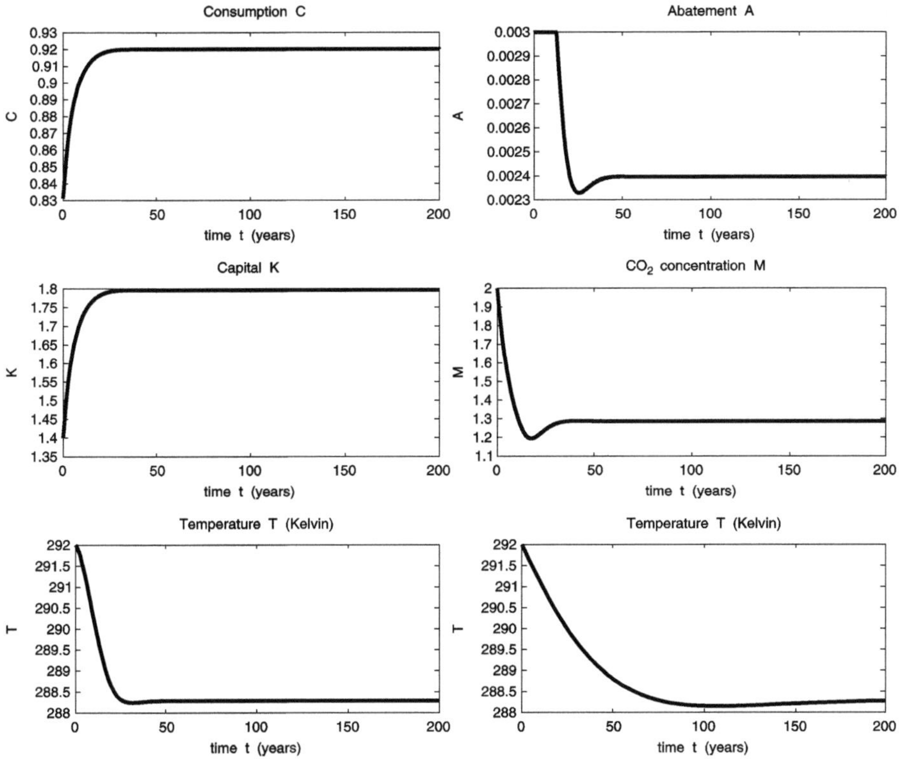

Fig. 16 *Finite horizon* $t_f = 200$, *social optimum with control* $u = (C, A)$. *Top row*: consumption C and abatement A. *Middle row*: capital K and CO_2 concentration M. *Bottom row*: (*left*) temperature T, (*right*) temperature T in infinite-horizon solution

6 Conclusion

In this paper, we study the canonical model of growth and climate change as put forward by Nordhaus' work (Nordhaus and Boyer 2000; Nordhaus 2008) and explore extensions of the basic model with respect to different scenarios. Policy options to mitigate climate change are often constrained by political events, lack of coalition formation, and the countries' political and economic means. In our paper, we explore a large number of scenarios of how mitigation policies could be pursued. We study the implication of infinite and finite horizon models, investigate the BAU scenario (business as usual scenario with low level abatement), and contrast it with an optimal abatement policy for infinite and finite horizon.

In finite-horizon scenarios, we explore the implications of terminal constraints of the state variable and consider the impacts of state constraints (such as CO_2 and temperature constraints) on abatement policies and consumption. Imposing such constraints allows us to find feasible control strategies for keeping the temperature and CO_2 concentration at low levels while preserving acceptable levels of consumption and capital. We also study another approach of keeping the temperature at a desirable level by putting suitable quadratic penalties on temperature deviations. The numerical analysis of these scenarios takes advantage of modern numerical techniques for solving constrained optimal control problems. In particular, the constrained scenarios allow us to explore the implications for mitigation policies arising from the Kyoto treaty (CO_2 constraint) and the Copenhagen agreement (temperature constraint). It is in this sense that we want to understand the exploration of our suggested different scenarios as guidance for different policy options.

References

Aseev, S. M., & Kryazhimskiy, A. V. (2004). The Pontryagin maximum principle and transversality condition for a class of optimal control problems with infinite time horizons. *SIAM Journal on Control and Optimization*, *43*, 1094–1119.

Aseev, S. M., & Kryazhimskiy, A. V. (2007). The Pontryagin maximum principle and economic growth. *Proceedings of the Steklov Institute of Mathematics*, *257*(1), 1–255.

Betts, J. T. (2010). *Advances in design and control. Practical methods for optimal control and estimation using nonlinear programming* (2nd edn.). Philadelphia: SIAM.

Büskens, C., & Maurer, H. (2000). SQP–methods for solving optimal control problems with control and state constraints: adjoint variables, sensitivity analysis and real–time control. *Journal of Computational and Applied Mathematics*, *120*, 85–108.

Fourer, R., Gay, D. M., & Kernighan, B. W. (1993). *AMPL: a modeling language for mathematical programming*. N. Scituate: Duxbury.

Greiner, A., Gruene, L., & Semmler, W. (2010). Growth and climate change: threshold and multiple equilibria. In J. Crespo Cuaresma, T. Palokangas & A. Tarasyev (Eds.), *Dynamic systems, economic growth, and the environment* (pp. 63–78). New York: Springer.

Hartl, R. F., Sethi, S. P., & Vickson, R. G. (1995). A survey of the maximum principles for optimal control problems with state constraints. *SIAM Review*, *37*, 181–218.

Hestenes, M. (1966). *Calculus of variations and optimal control theory*. New York: Wiley.

Keller, K., Bolkerb, M. B., & Bradford, D. F. (2004). Uncertain climate thresholds and optimal economic growth. *Journal of Environmental Economics and Management*, *48*(1), 723–741.

Keller, K., Tan, K., Morel, F. M., & Bradford, D. F. (2000). Preserving the ocean circulation: implications for the climate policy. *Climate Change*, *47*, 17–43.

Kunkel, P., & von dem Hagen, O. (2000). Numerical solution of infinite-horizon optimal control problems. *Computational Economics*, *16*, 189–205.

Lykina, V. (2010). *Beiträge zur Theorie der Optimalsteuerungsprobleme mit unendlichem Zeithorizont*. Dissertation, Bandenburgische Technische Universität Cottbus. Cottbus, Germany.

Lykina, V., Pickenhain, S., & Wagner, M. (2008). Different interpretations of the improper integral objective in an infinite horizon control problem. *Journal of Mathematical Analysis and Applications*, *340*, 498–510.

Maurer, H. (1979). *On the minimum principle for optimal control problems with state constraints* (Report no. 41). Münster, Germany: Rechenzentrum der Universität Münster.

Michel, P. (1982). On the transversality conditions in infinite horizon optimal control problems. *Econometrica*, *50*, 975–985.

Nordhaus, W. (2008). *The question of balance*. New Haven: Yale University Press.

Nordhaus, W. D., & Boyer, J. (2000). *Warming the world. Economic models of global warming*. Cambridge: MIT Press.

Pontryagin, L. S., Boltyanski, V. G., Gramkrelidze, R. V., & Miscenko, E. F. (1964). *The mathematical theory of optimal processes*. Moscow: Fitzmatgiz. English translation: New York: Pergamon.

Roedel, W. (2011). Private communication. June 2011.

Roedel, W., & Wagner, T. (2011). *Physik unserer Umwelt: Die Atmosphäre*. Berlin: Springer.
Seierstadt, A., & Sydsaeter, K. (1987). *Optimal control theory with economic applications*. Amsterdam: North-Holland.
Sethi, S. P., & Thompson, G. L. (2000). *Optimal control theory: applications to management science and economics* (2nd edn.). Norwell: Kluwer Academic.
Wächter, A., & Biegler, L. T. (2006). On the implementation of an interior-point filter line-search algorithm for large-scale nonlinear programming. *Mathematical Programming*, *106*(1), 25–57; cf. Ipopt home page (C. Laird and A. Wächter): https://projects.coin-or.org/Ipopt.

Part III
Dynamics of Environmental Policy with an Oligopoly

Market Power, Resource Extraction and Pollution: Some Paradoxes and a Unified View

Luca Lambertini and George Leitmann

1 Introduction

The conflict between individual incentives and the preservation of the environment and natural resources, and the associated market failures, are well known since Gordon (1954) and Hardin's (1968) *tragedy of the commons*. In the subsequent decades, the economic literature has produced countless contributions concerning either the exploitation of natural resources or the environmental externalities generated by industrial activities, but rarely—if ever—both at the same time, although the interplay between growth and the environment and the sustainability of our economic system are both generally viewed as a circular model with feedback effects.[1]

This partial approach to a single side of the problem at a time is quite common in both static and dynamic applications of oligopoly theory to environmental or resource economics. Some recurrent themes emerging from this strand of literature can be quickly recollected so as to fix ideas. A cornerstone of the discussion is the market failure associated with external effects:

- Firms do not internalise environmental externalities, and therefore will not spontaneously invest in green technologies. This prompts the design of Pigouvian taxation to supply the proper R&D incentives to firms.[2]

[1] This view is so largely shared in the profession, that it appears regularly in the introductory chapters of textbook at any level (see, e.g., Pearce and Turner 1989; Tisdell 2009; and Anderson 2010).

[2] See Downing and White (1986), Milliman and Prince (1989), Karp and Livernois (1994), Chiou and Hu (2001) and Poyago-Theotoky (2007), inter alia.

L. Lambertini (✉)
Department of Economics, University of Bologna, Strada Maggiore 45, Bologna 40125, Italy
e-mail: luca.lambertini@unibo.it

G. Leitmann
College of Engineering, University of California at Berkeley, Berkeley, CA 94720, USA
e-mail: gleit@berkeley.edu

J. Crespo Cuaresma et al. (eds.), *Green Growth and Sustainable Development*, Dynamic Modeling and Econometrics in Economics and Finance 14,
DOI 10.1007/978-3-642-34354-4_7,

- For analogous reasons, firms may overexploit natural resources, renewable or not.[3]

Another critical point concerns the pros and cons of any variation in industry output. Any industrial economist would agree that market power is detrimental to welfare because of high prices and low output levels. However, expanding output puts additional pressure upon the environment and the stock of natural resources. This reveals the presence of a tradeoff between the static efficiency usually associated with marginal cost pricing and the dynamic efficiency one should refer to in connection with the long-run sustainability of the current economic system.

Here we set out to revisit these issues through a differential game approach which, unlike the majority—if not all—of the existing contributions in this field,[4] explicitly incorporates the exploitation of natural resources and the environmental consequences of industrial production in a single setting, to investigate their interplay in such a way that the tragedy of commons can indeed be reinterpreted as simultaneously involving the depletion of natural resources and an undesirable impact on the environment, these two facts being (i) two sides of the same coin and (ii) both driven by the selfishness implicit in firms' pure profit incentives. To put it differently, we propose a way of modelling environmental externalities as a direct consequence of the tragedy of commons associated with the exploitation of natural resources.

Although being not properly a general equilibrium one, our setup draws explicitly the endogenous link between the dynamics of resource exploitation and environmental externalities, to show that (i) pure profit incentives can indeed give rise to investments in green technologies which are ruled out in the conventional approach, and (ii) competition may (but not necessarily does) exert positive long-run welfare effects, although with mixed feelings, as a cleaner environment is accompanied by a lower residual stock of natural resources. Our procedure will be the following. We will set out with the illustration of simple setups alternatively accounting for the presence of either natural resources or pollution, in open-loop games in which firms control output levels. Then we will enrich the picture introducing a simple production function accounting for the fact that the natural resource enters the productive activities of firms as an input, and pollution may be subject to Pigouvian taxation which can be used as an incentive for green R&D. Then we will lay out a comprehensive model capturing the interplay between the output and R&D decisions of firms on one side and the preservation of natural resources and the environment on the other. Whenever appropriate, we will also dwell on the optimal industry structure (i.e., the number of firms) in the commons, an issue that has received a considerable amount of attention in the early debate on common property in oligopoly (on this,

---

[3]Classical contributions in this vein are those of Clark and Munro (1975) and Levhari and Mirman (1980). For a model of international trade with natural resource extraction, see Copeland and Taylor (2009), inter alia. An overview of the debate is in Long (2010).

[4]See, for instance, Chap. 12 in Dockner et al. (2000) and Chaps. 2–3 in Long (2010), offering surveys of dynamic games in which environmental externalities and resource extraction are treated separately.

see Cornes and Sandler 1983; Cornes et al. 1986; and Mason and Polasky 1997, inter alia). In particular, the full model involving an endogenous link between resource extraction and pollution shows that the socially efficient industry structure is sensitive to the environment's natural recycling capabilities. It turns out that monopoly is socially optimal if either the rate of reproduction of the natural resource is large enough or the natural rate of absorption of polluting emissions is low enough, as in both cases the usual output restriction associated with monopoly power reduces the environmental impact of production more than consumer surplus.

The remainder of the paper is structured as follows. In Sect. 2, we offer a step-by-step reconstruction of the standard approach, whereby resource extraction and pollution are studied in isolation from each other. Then, in Sect. 3, we propose a fully fledged model taking into account the interplay between state variables. Concluding remarks are in Sect. 4.

2 Preliminaries: The Standard Approach

Here we summarise the acquired wisdom based on previous literature, where either pollution or the exploitation of natural resources have been treated in isolation from one another. Throughout, we will consider an industrial sector existing over continuous time $t \in [0, \infty)$.

2.1 Natural Resources I

The simplest model of the interplay between profit incentives and resource extraction is the following. The market is supplied by n firms offering a homogeneous good produced through a renewable natural resource (say, forestry) to deliver an intermediate or final commodity (say, timber or paper) to consumers. The market demand for the final good is

$$p(t) = a - Q(t), \quad Q(t) = \sum_{i=1}^{n} q_i(t), \tag{1}$$

$q_i(t)$ being the instantaneous output of firm $i = 1, 2, 3, \ldots, n$. Therefore, the game features n controls, $\mathbf{q} = (q_1, q_2, \ldots, q_n)$, one for each player. All firms share a symmetric technology with the same marginal cost c for extraction and production, giving rise to a cost function $C_i(t) = cq_i(t)$, with $a > c > 0$. This imposes the constraint $Q(t) \in [0, a - c]$ at any time t. Additionally, the difference between reservation price and marginal cost, $a - c$, is assumed to be large enough to ensure the non-negativity of controls at all times during this and the subsequent versions of the game. The instantaneous profit function of firm i is $\pi_i(t) = (p(t) - c)q_i(t)$.

The only state variable appearing in this version of the model is the stock of the natural resource $x(t) \geq 0$, which evolves over time according to the following state equation:[5]

$$\dot{x}(t) = \eta x(t) - \sum_{i=1}^{n} q_i(t) \tag{2}$$

where $\eta > 0$ is the constant rate of reproduction.

The game is non-cooperative and simultaneous play takes place at any instant. The individual firm has to

$$\max_{q_i(t)} \int_0^{\infty} \pi_i(t) e^{-\rho t} dt \tag{3}$$

subject to the dynamic constraints (2), the initial condition $x(0) = x_0 \geq n(a-c)/(n+1)/\eta$,[6] and the appropriate transversality condition. The discount rate $\rho > 0$ is constant and common to all firms.

Firm i's Hamiltonian function is

$$\mathcal{H}_i(t) = e^{-\rho t} \left\{ \left(a - q_i(t) - \sum_{j \neq i} q_j(t) - c \right) q_i(t) + \lambda_i(t) \left[\eta x(t) - \sum_{i=1}^{n} q_i(t) \right] \right\} \tag{4}$$

in which $\lambda_i(t) = e^{\rho t} \gamma_i(t)$ is the co-state variable (in current value) associated with the dynamics of the state. This being a linear state game, the open-loop solution is subgame perfect (or equivalently, strongly time consistent).

The necessary conditions are[7]

$$\frac{\partial \mathcal{H}_i}{\partial q_i} = e^{-\rho t} \left(a - c - 2q_i - \sum_{j \neq i} q_j - \lambda_i \right) = 0 \tag{5}$$

$$\dot{\lambda}_i = (\rho - \eta) \lambda_i \tag{6}$$

---

[5]We could model the state equation as

$$\dot{x}(t) = \eta x(t) - \nu \sum_{i=1}^{n} q_i(t)$$

with $\nu \in (0, 1]$. This, however, would not modify significantly the qualitative predictions of the our analysis. Therefore, we have imposed $\nu = 1$ to restrict the set of parameters.

[6]Taking $x_0 > 0$ as the initial condition does not ensure the sustainability of extraction activities over $t \in [0, \infty)$ as the stock $x(t)$ would become nil in finite time.

[7]Henceforth, we omit the time argument for the sake of brevity. Mangasarian's (1966) and Arrow's (1968) sufficiency conditions are also satisfied. They are also omitted for the sake of brevity.

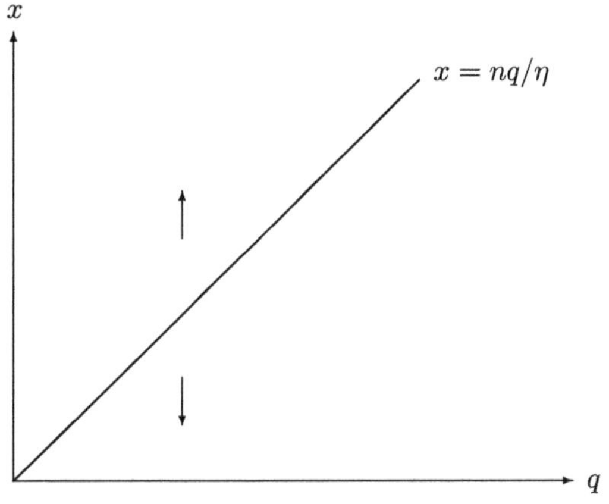

Fig. 1 Phase diagram in the (q, x) space

together with the transversality conditions $\lim_{t\to\infty} \gamma_i x = \lim_{t\to\infty} e^{-\rho t}\lambda_i x = 0$. Now observe that (6) admits the solution $\lambda_i = 0$ at all times, whereby the first order condition on the output level yields the static Cournot-Nash solution $q_i = q^N = (a-c)/(n+1)$ for all i.[8] Then, imposing stationarity on $\dot{x}$, one gets

$$x^* = \frac{nq^N}{\eta} = \frac{n(a-c)}{(n+1)\eta}, \tag{7}$$

with $\partial x^*/\partial n > 0$, which seems to indicate that increasing competition has positive consequences for the preservation of the natural resource in the long run.

Since the output is constant throughout the game, the dynamics of the model reduces to (2), as depicted in Fig. 1. As $\dot{x} \gtreqless 0$ for all $x \gtreqless nq/\eta$, it appears that perturbations could cause the system to diverge, in either direction. This—which, literally, is a technical feature of the model—lends itself to an intuitive interpretation, which can be spelled out in the following terms. Firms' myopic behaviour, dictated by pure profit incentives, may indeed cause the delicate equilibrium between the economic system and the environment to collapse (of course x could indeed experience a limitless growth, but casual observation suggests the opposite).

The bottom line of this exercise is well known, as it states that firms replicate forever the equilibrium of the static Cournot game without internalising the effects of production on the existing amount of the natural resource.[9] The industry output is therefore increasing with n, and this causes, at any time t, an increase in the

[8]Throughout the paper, we shall use superscript N to identify the Nash solution, while starred values will indicate steady state magnitudes of states and controls. Note that in the present section the optimal output is stationary as firms do not take into account the consequences of their behaviour on the stock of resources. Hence, here the steady state output is the same as the Nash equilibrium output at any time t. The same applies to the model illustrated in Sect. 2.3.

[9]This is the reason why we have taken the initial stock to be at least as large as the Cournot-Nash industry output.

rate of extraction as compared to pure monopoly, due to the output restriction that is usually associated with monopolistic sectors, as compared to any even slightly more competitive industries. On the other hand, any increase in output lowers market price and brings about an increase in consumer surplus $CS^* = (Q^*)^2/2$. The balance between these effects is captured by the net effect of a change in n on the social welfare function[10]

$$SW^* = n\pi^* + CS^* + x^* \tag{8}$$

with

$$\frac{\partial SW^*}{\partial n} = \frac{(a-c)[n+1+\eta(a-c)]}{\eta(n+1)^3} > 0 \tag{9}$$

for all n. That is, competition is promoting social welfare, notwithstanding the fact that it involves a higher exploitation rate at any time. The next step consists in taking into consideration a slightly richer version of the same problem, which explicitly acknowledges the presence of an endogenous link between the natural resource and the output via a simple production function.

2.2 *Natural Resources II*

Define now the individual output of firm i as $q_i = b_i x$, b_i being the instantaneous rate at which firm i extract the resource and uses it in the production of the intermediate or final good. This can be thought of as a production function operating at constant returns to scale, using the natural resource as the only relevant input. We shall see in the remainder of the section that this seemingly simple transformation indeed has relevant consequences on our understanding and interpretation of the problem at hand. As in the previous version, we have n controls, $\mathbf{b} = (b_1, b_2, \ldots, b_n)$, one for each player, while there is a single state, x, whose dynamics is now

$$\dot{x} = \left(\eta - \sum_{i=1}^{n} b_i\right)x, \tag{10}$$

so that the insertion of a simple linear technology in the model establishes a multiplicative effect between state and control in the state equation which was altogether absent in the previous version. The remainder of the setup is unmodified, so the Hamiltonian of firm i is

---

[10]Note that the amount of natural resource enters the social welfare function with a weight equal to one, i.e., the same attached to industry profits and consumer surplus. The ongoing debate on this point has not yet produced a unanymous view (see, e.g., Chap. 5 in Stern 2009). The need for guaranteeing the prosperity of future generations suggests that one should attach to the preservation of natural resources at least the same importance as traditional economic indicators strictly related to production and consumption.

$$\mathcal{H}_i = e^{-\rho t}\left[\left(a - b_i x - x\sum_{j\neq i} b_j - c\right)b_i x + \lambda_i\left(\eta - \sum_{i=1}^{n} b_i\right)x\right]. \tag{11}$$

The game is thus no longer a linear state one,[11] but for the sake of comparability we stick to the open-loop solution, requiring the following necessary conditions:

$$\frac{\partial \mathcal{H}_i}{\partial b_i} = e^{-\rho t} x\left(a - c - 2b_i x - x\sum_{j\neq i} b_j - \lambda_i\right) = 0 \tag{12}$$

$$\dot{\lambda}_i = \left(\rho - \eta + \sum_{i=1}^{n} b_i\right)\lambda_i - b_i\left[a - c - 2x\left(\sum_{j\neq i} b_j + 2b_i\right)\right], \tag{13}$$

while the transversality conditions are $\lim_{t\to\infty} e^{-\rho t}\lambda_i x = 0$.

To simplify calculations, henceforth we impose symmetry on controls and co-states, $b_i = b_j = b$ and $\lambda_i = \lambda_j = \lambda$ for all i, j. From (12) we obtain

$$b = \max\left\{\frac{a - c - \lambda}{(n+1)x}, 0\right\} \tag{14}$$

and, if $b > 0$,

$$\lambda = a - c - b(n+1)x. \tag{15}$$

Then, employing (10), we can write

$$\dot{b} = -\frac{x\dot{\lambda} + (a - c - \lambda)\dot{x}}{(n+1)x^2}. \tag{16}$$

which, using (13)–(15), can be rewritten as

$$\dot{b} = \frac{(a-c)[\eta - \rho - b(n-1)] + b[\rho + (2nb + \rho)n - 2(n+1)\eta]x}{(n+1)x^2} \tag{17}$$

Imposing stationarity on state and control, we identify the unique open-loop steady state equilibrium, where:[12]

$$b_i^* = b^* = \frac{\eta}{n} \quad \forall i; \qquad x^* = \frac{(a-c)(\eta - n\rho)}{\eta[2\eta - (n+1)\rho]}. \tag{18}$$

Note that $x^* > 0$ for all

[11] Additionally, note that it is not defined in linear-quadratic form. Consequently, we have no obvious conjecture as to the form of the value function.

[12] The corresponding value of the co-state variable at the steady state equilibrium is

$$\lambda^* = \frac{(a-c)(n-1)\eta}{n[2\eta - (n+1)\rho]}$$

and the transversality condition $\lim_{t\to\infty} e^{-\rho t}\lambda_i x = 0$ is met thanks to exponential discounting.

$$\rho \in \left(0, \frac{\eta}{n}\right) \quad \text{and} \quad \rho > \frac{2\eta}{n+1}. \tag{19}$$

The same obviously holds for the individual output $q^* = b^* x^*$ as well as for the industry output $Q^* = nq^*$. Steady state individual profits are

$$\pi^* = \frac{(a-c)^2(\eta-\rho)(\eta-n\rho)}{n[2\eta-(n+1)\rho]^2} > 0 \quad \forall \rho \in \left(0, \frac{\eta}{n}\right). \tag{20}$$

Hence, the survival of firms at the steady state requires indeed $\rho \in (0, \eta/n)$. Now observe that

$$\frac{\partial x^*}{\partial n} = \frac{(a-c)(\rho-\eta)}{\eta[2\eta-(n+1)\rho]^2} < 0 \tag{21}$$

for all $\rho \in (0, \eta)$. In view of the previous result, this is surely the case for any $n \geq 1$. This contradicts the result of the previous—and simpler—version of this problem, as now it appears that a more intense competition throughout the game leads to a lower amount of natural resource left over at the steady state.

Now reconsider the social welfare function, defined as in (8). Its partial derivative w.r.t. n is

$$\frac{\partial SW^*}{\partial n} = \frac{(a-c)(\eta-\rho)[\rho(n+1+\eta(a-c)) - \eta(2+\eta(a-c))]\rho}{\eta[2\eta-(n+1)\rho]^3} \tag{22}$$

which is positive iff

$$\rho > \frac{\eta(2+\eta(a-c))}{n+1+\eta(a-c)}. \tag{23}$$

However,

$$\frac{\eta(2+\eta(a-c))}{n+1+\eta(a-c)} - \frac{2\eta}{n+1} = \frac{\eta^2(a-c)(n-1)}{[n+1+\eta(a-c)](n+1)} > 0 \quad \forall n \geq 2. \tag{24}$$

Hence, the equilibrium level of social welfare monotonically decreases with n once the interplay between the natural resource and the firms' output has been duly accounted for, although admittedly in a very simple manner.[13] Once again, the sign of the partial derivative (22) is reversed as compared to the previous setup. The foregoing discussion can be summarised in

Proposition 1 *Endogenising the technological link between the exploitation of the natural resource and the intermediate or final output of the industry singles out the negative effect of an increase in the intensity of competition on the resulting equilibrium level of social welfare.*

[13]This result relates the long-run effects of increasing the population of firms with the discount rate. This aspect is a crucial feature of an ongoing debate concerning the need for applying low discount rates to the well-being of future generation (see Stern 2007; and Weitzman 2007, inter alia).

We can now move on to the stability analysis of the dynamic system formed by (10) and (17). This is carried out by investigating the features of the Jacobian matrix:

$$J = \begin{bmatrix} \frac{\partial \dot{x}}{\partial x} & \frac{\partial \dot{x}}{\partial b} \\ \frac{\partial \dot{b}}{\partial x} & \frac{\partial \dot{b}}{\partial b} \end{bmatrix}$$

whose determinant, in correspondence of the unique steady state point, is

$$\Delta(J) = \frac{\eta[\rho(n+1) - 2\eta]}{n+1} < 0 \quad \forall \rho \in \left(0, \frac{\eta}{n}\right). \tag{25}$$

This suffices to prove:

Proposition 2 *The unique steady state equilibrium* (x^*, b^*) *is a saddle point.*

2.3 Pollution I

Let us now turn to an alternative scenario where natural resources are left out of the picture and the focus is on the environmental consequences of production (or consumption). Still, we consider the same n-firm oligopoly offering a homogeneous good, which now generates an undesirable environmental externality. Let $s(t) \geq 0$ be the stock of environmental pollution at any instant. We assume that pollution follows the dynamic equation:

$$\dot{s} = z \sum_{i=1}^{n} q_i - \sum_{i=1}^{n} k_i - \delta s \tag{26}$$

that is, it increases with the industry output level $Q = \sum_{i=1}^{n} q_i$ at a constant rate $z > 0$, while it decreases with the industry's green R&D investments $K = \sum_{i=1}^{n} k_i$, k_i being the instantaneous R&D effort of firm i at the cost $\Gamma_i = vk_i^2$, parameter $v > 0$ measuring the marginal cost of R&D. Pollution by itself diminishes at the constant rate $\delta \geq 0$. Hence, we are considering a single state, s and $2n$ controls, $\mathbf{q} = (q_1, q_2, \ldots, q_n)$ and $\mathbf{k} = (k_1, k_2, \ldots, k_n)$, two for each player.

The market demand function is $p = a - Q$, while firm i's production involves a cost function $C_i(q_i) = cq_i$, $c > 0$. The Hamiltonian of firm i is

$$\mathcal{H}_i = e^{-\rho t}\left[\left(a - q_i - \sum_{j \neq i} q_j - c\right)q_i - vk_i^2 + \mu_i\left(z\sum_{i=1}^{n} q_i - \sum_{i=1}^{n} k_i - \delta s\right)\right] \tag{27}$$

where $\mu_i = e^{\rho t}\varpi_i$ is the co-state variable (in current value) associated with s. Strategic interaction is simultaneous, and we shall focus on the open-loop non-cooperative Nash solution. This game is a linear state one, so that (i) the open-loop Nash equilibrium is subgame perfect, and (ii) we may anticipate that unregulated firms will

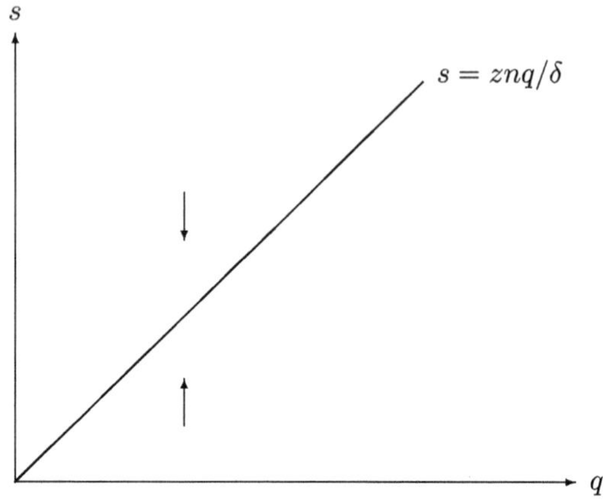

Fig. 2 Phase diagram in the (q, s) space

never spontaneously internalise the environmental consequences of their productive activities at any time during the game. This ultimately implies that they will not invest in R&D, as can be ascertained through a quick examination of the necessary conditions:

$$\frac{\partial \mathcal{H}_i}{\partial q_i} = e^{-\rho t}\left(a - c - 2q_i - \sum_{j \neq i} q_j - z\mu_i\right) = 0 \tag{28}$$

$$\frac{\partial \mathcal{H}_i}{\partial k_i} = -e^{-\rho t}(2vk_i + \mu_i) = 0 \tag{29}$$

$$\dot{\mu}_i = (\delta + \rho)\mu_i \tag{30}$$

Since (30) admits the solution $\mu_i = 0$ at all times,[14] this immediately entails that $k_i^N = 0$ and $q_i^N = (a - c)/(n + 1)$ at any time t. The transversality condition $\lim_{t\to\infty} \varpi_i s = \lim_{t\to\infty} e^{-\rho t}\mu_i s = 0$ is also met. The resulting level of pollution at the steady state is $s^* = n(a - c)z/[(n + 1)\delta]$, with $\partial s^*/\partial n > 0$.

As in Sect. 2.1, again here the system dynamics reduces to the state equation (26). The crucial difference is that here a stability property emerges, since $\dot{s} \gtrless 0$ for all $s \lessgtr znq/\delta$. This is depicted in Fig. 2.

2.4 Pollution II

As is well known, the way out of this *impasse* consists in introducing a Pigouvian taxation/subsidization $\mathcal{P}$ proportional to the stock of pollution, say, $\mathcal{P} = \theta s$, on all firms alike (see, e.g., Benchekroun and Long 1998, 2002), θ being the tax or subsidy rate, which we will take to be constant throughout the game, for reasons that will

[14]Note that $\mu_i = 0$ suffices to ensure that the transversality condition $\lim_{t\to\infty} \mu_i s = 0$ be satisfied.

be clarified below.[15] This entails that instantaneous per firm profits are $\pi_i = (p - c)q_i - vk_i^2 - \theta s$, and transforms the Hamiltonian of firm i into the following:

$$\mathcal{H}_i = e^{-\rho t}\left[\left(a - q_i - \sum_{j\neq i} q_j - c\right)q_i - vk_i^2 - \theta s \right.$$
$$\left. + \mu_i\left(z\sum_{i=1}^{n} q_i - \sum_{i=1}^{n} k_i - \delta s\right)\right]. \tag{31}$$

As before, firms play simultaneously and non-cooperatively at all times, taking now as given the Pigouvian policy set by the government. Observe that a direct consequence of the presence of regulation is that

$$\dot{\mu}_i = (\delta + \rho)\mu_i + \theta \tag{32}$$

which does not admit the nil solution any more and therefore opens the way for positive R&D investments and also influences firms' output decisions:

$$q_i = q^* = \frac{a - c - 2vzk^*}{n+1}; \qquad k_i = k^* = \frac{\theta}{2(\delta + \rho)v} \quad \forall i \tag{33}$$

with $q^* > 0$ provided $a - c > z\theta/(\delta + \rho)$. Output q^* is decreasing with k^* and the steady state R&D effort k^* is increasing with θ, in such a way that—if firms are being taxed, i.e., for all $\theta > 0$—R&D efforts are positive and the industry output is lower than in the unregulated case (the opposite holds of course if firms are subsidised, which happens for $\theta < 0$). The resulting level of pollution at the steady state equilibrium is

$$s^* = \frac{n[2v(a-c)(\delta+\rho)z - \theta(n+1+2vz^2)]}{2\delta(n+1)(\delta+\rho)v}. \tag{34}$$

The stability analysis is readily dealt with, as at any time the optimal individual quantity is indeed as in (33), so that dynamic system involves the behaviour of pollution and green R&D only. Hence, focussing on the Jacobian matrix:

$$J = \begin{bmatrix} \dfrac{\partial \dot{s}}{\partial s} & \dfrac{\partial \dot{s}}{\partial k} \\[2ex] \dfrac{\partial \dot{k}}{\partial s} & \dfrac{\partial \dot{k}}{\partial k} \end{bmatrix}$$

one can write its determinant, $\Delta(J) = -\delta(\delta + \rho)$, which is always negative. This implies:

Proposition 3 *The unique steady state equilibrium* (s^*, k^*, q^*) *is a saddle point.*

[15] Therefore, we have a single state, s and $2n + 1$ controls, i.e., $\mathbf{q} = (q_1, q_2, \ldots, q_n)$ and $\mathbf{k} = (k_1, k_2, \ldots, k_n)$, two for each firm, and the Pigouvian policy rate θ for the government.

Then the question arises as to how the government should choose the optimal θ. Since θ appears in the firms' first order conditions, the open-loop solution of the Stackelberg game with the government leading would obviously be subject to a problem of time inconsistency. To avoid it, one may either solve the Stackelberg game via a degenerate Markov approach (see Chap. 5 in Dockner et al. 2000), or simply calculate the value of θ maximising the steady state level of social welfare[16] defined as

$$SW^* = n\pi^* + CS^* - s^* + n\mathcal{P} = n\pi^* + CS^* - s^*(1 - n\theta), \tag{35}$$

since the tax revenues $n\mathcal{P} = n\theta s$ are assumed to be redistributed across consumers as windfall money. This requires solving

$$\frac{\partial SW^*}{\partial \theta} = 0 \tag{36}$$

yielding

$$\theta^* = \frac{(\delta + \rho)[(n+1)(n+1+2vz^2) - 2\delta(a-c)vz]}{\delta[n^2 + 1 + 2(1+vz^2)n]} \tag{37}$$

which may take either positive or negative values depending on the relative size of parameters $\{a, c, n, v, z, \delta, \rho\}$, in particular n and $a - c$. We have that

$$SW^*\big|_{\theta=\theta^*} > SW^*\big|_{\theta=0} \tag{38}$$

always, while

$$\operatorname{sign} s^*\big|_{\theta=\theta^*} - s^*\big|_{\theta=0} = \operatorname{sign}\big[2\delta(a-c)vz - (n+1)\big(n+1+2vz^2\big)\big] \tag{39}$$

which delivers the following:

Proposition 4 *For all*

$$a - c \in \left(0, \frac{(n+1)(n+1+2vz^2)}{2\delta vz}\right),$$

$\theta > 0$ *and* (i) $SW^*|_{\theta=\theta^*} > SW^*|_{\theta=0}$; (ii) $s^*|_{\theta=\theta^*} < s^*|_{\theta=0}$.

For all

$$a - c > \frac{(n+1)(n+1+2vz^2)}{2\delta vz},$$

$\theta < 0$ *and* (i) $SW^*|_{\theta=\theta^*} > SW^*|_{\theta=0}$; (ii) $s^*|_{\theta=\theta^*} > s^*|_{\theta=0}$.

The second claim appearing in the above Proposition states that, if either the reservation price a is high enough or the marginal cost c is sufficiently low, the Pigouvian policy takes the form of a subsidy leading to a level of pollution higher

[16]This is the route taken by Benchekroun and Long (1998, 2002).

than it would be without regulation. The increase in welfare, which obtains irrespective of whether firms are being taxed or subsidised, obtains because of the output expansion that is brought about by subsidization and ultimately increases consumer surplus. This is a direct consequence of the aforementioned tradeoff between the price effect and the external effect which, provided the market is affluent enough, paradoxically induces the regulator to opt for a higher consumer surplus even though this entails a larger amount of pollution.[17]

It can be shown that $\partial SW^*|_{\theta=\theta^*}/\partial n > 0$ always.[18] The analysis of the effects of a change in n on industry profits can only be carried out numerically, revealing that the industry concentration which maximises collective profits is increasing in δ. To see this, we fix $a - c = 1$, $v = 3$, $z = 1$ and $\rho = 1/5$, and solve $\partial n\pi^*|_{\theta=\theta^*}/\partial n = 0$ for different values of δ, obtaining (n is rounded to the lower integer):

$$\begin{aligned} n &= 58 \quad \text{for } \delta = \frac{1}{5} \\ n &= 73 \quad \text{for } \delta = \frac{1}{4} \\ n &= 99 \quad \text{for } \delta = \frac{1}{3} \\ n &= 149 \quad \text{for } \delta = \frac{1}{2} \\ n &= 224 \quad \text{for } \delta = \frac{3}{4} \\ n &= 239 \quad \text{for } \delta = \frac{4}{5} \end{aligned} \tag{40}$$

That is,

Remark 5 The higher the environment's degree of efficiency in recycling pollution, the larger is the population of firms maximising industry profits at the steady state equilibrium in which a benevolent regulator adopts the socially optimal Pigouvian policy.

3 The Full Model

We are now ready to investigate a full-fledged model in which the natural resource enters explicitly in the production function of the intermediate or final output, and

---

[17]It is worth noting that this mechanism would still exist in a simpler version of this setup, without R&D investments. This is due to the fact that the conflict between two equally desirable objectives (lowering the price and reducing pollution) is entirely inherent in production decisions only.

[18]The proof, trivial but lengthy, is omitted for brevity. It is however available from the authors upon request.

productive activities generate a negative environmental externality. To do so, we modify the state equations as follows.

We pose that pollution evolves according to the following equation:

$$\dot{s} = zx \sum_{i=1}^{n} b_i - \sum_{i=1}^{n} k_i - \delta x, \tag{41}$$

where the only detail that has changed as compared to the previous version is that the environment is being cleaned at a rate $\delta \geq 0$ by the existing amount of natural resource.[19]

The dynamics of the natural resource is

$$\dot{x} = \left(\eta - \sum_{i=1}^{n} b_i \right) x - s, \tag{42}$$

in which, it should be noted, the stock of pollution enters negatively.

All of the control variables have been already defined. Thus, the present game features two state variables, s and x, and $2n$ controls, $\mathbf{q} = (q_1, q_2, \ldots, q_n)$ and $\mathbf{k} = (k_1, k_2, \ldots, k_n)$, two for each player. We disregard the possibility of regulation through a Pigouvian policy for reasons that will become apparent below. The instantaneous profit function of firm i is

$$\pi_i = (p - c) b_i x - v k_i^2 \tag{43}$$

so that the individual firm must

$$\max_{b_i, k_i} \int_0^{\infty} \pi_i e^{-\rho t} dt \tag{44}$$

subject to the dynamic constraints (41)–(42), initial conditions $s(0) = s_0 > 0$ and $x(0) = x_0 > s/(\eta - \sum_{i=1}^{n} b_i)$, and the appropriate transversality conditions. Once again, we solve the game under open-loop information. The firm's Hamiltonian function is

$$\begin{aligned} \mathcal{H}_i = e^{-\rho t} \Bigg\{ & \left(a - b_i x - x \sum_{j \neq i} b_j - c \right) b_i x - v k_i^2 \\ & + \varphi_i \left(zx \sum_{i=1}^{n} b_i - \sum_{i=1}^{n} k_i - \delta x \right) + \psi_i \left[\left(\eta - \sum_{i=1}^{n} b_i \right) x - s \right] \Bigg\}, \end{aligned} \tag{45}$$

variables $\varphi_i = e^{\rho t} \zeta_i$ and $\psi_i = e^{\rho t} \varkappa_i$ being the co-states (in current value) associated with $s(t)$ and $x(t)$, respectively. The maximization of (45) requires meeting the

[19] This applies to rain forests and the oceans absorbing CO_2 emissions, while it does not apply to other natural resources, like the stock of fish. On the contrary, the latter is negatively affected by pollution (as specified in (2), (10) and (42)).

following set of necessary conditions:

$$\frac{\partial \mathcal{H}_i}{\partial b_i} = e^{-\rho t} x \left(a - c - 2b_i x - x \sum_{j \neq i} b_j + z\varphi_i - \psi_i \right) = 0 \tag{46}$$

$$\frac{\partial \mathcal{H}_i}{\partial k_i} = -e^{-\rho t} (2vk_i + \varphi_i) = 0 \tag{47}$$

$$\dot{\varphi}_i = \rho \varphi_i + \psi_i \tag{48}$$

$$\dot{\psi} = (\rho - \eta + B)\psi + (\delta - zB)\varphi_i - b_i(a - c - 2xB) \tag{49}$$

with $B \equiv \sum_{i=1}^{n} b_i$ measuring the industry extraction rate. The associated transversality conditions are

$$\lim_{t \to \infty} e^{-\rho t} \varphi_i s = 0; \qquad \lim_{t \to \infty} e^{-\rho t} \psi_i x = 0 \quad \forall i. \tag{50}$$

Observe that (47) entails that, if the individual firm attaches a negative shadow value to the environmental damage, i.e. $\varphi_i < 0$, the resulting instantaneous green R&D effort $k_i = -\varphi_i/(2v)$ is positive. As we know, this usually does not happen if firms are unregulated (see Sects. 2.3–2.4). We are about to see that allowing for an endogenous interaction between natural resources and the environmental consequences of production modifies this crucial aspect, even in the absence of any intervention by a policy maker.

From (46), we obtain

$$\psi_i = a - c - 2b_i x - x \sum_{j \neq i} b_j + z\varphi_i \tag{51}$$

and

$$\dot{b}_i = \frac{x(z\dot{\varphi}_i - \dot{\psi}_i - x \sum_{j \neq i} \dot{b}_j) - (a - c + z\varphi_i - \psi_i)\dot{x}}{2x^2} \tag{52}$$

while from (47) we get

$$\varphi_i = -2bk_i \tag{53}$$

and therefore also

$$\dot{k}_i = -\frac{\dot{\varphi}_i}{2b} = -\frac{\rho\varphi_i + \psi_i}{2b}. \tag{54}$$

Before proceeding any further, it is worth noting that (47) and (54) jointly imply:

Lemma 6 *Given $\rho > 0$, any triple $\{\rho, \varphi_i, \psi_i\}$ such that $\varphi_i < 0$ and $\psi_i < -\rho\varphi_i$ suffices to ensure the presence of positive and increasing R&D efforts during the game.*

Namely, the interplay between states may indeed act as a substitute for regulation, the reason being that the source of the environmental externality (which, in the traditional approach to the matter, is deemed irrelevant to pure profit-seeking firms) is to be traced back to the exploitation of a natural resource for production whose extraction and use is costly to the firm itself. This endogenous link does not turn the firm into an altruistic agent, but forces it to become aware of the social cost of its own activities as they exert an impact on the technological cost of production. Consequently, the firm may find it profitable to search for cleaner technologies by virtue of its own profit incentives only.

Substituting (48), (49), (51) and (53) into (52) and (54), and imposing symmetry across firms, we can write the control equations:

$$\begin{aligned}\dot{b} &= \big\{(a-c)\big[\eta-\rho+z-b(n-1)\big]+2vk\big[\delta+z\big(b(n-1)-\eta-z\big)\big]\\&\quad+b(n+1)\big[s-x\big(b-\rho+z-2(nb-\eta)\big)\big]\big\}/\big[(n+1)x\big] \qquad (55)\\ \dot{k} &= \frac{b(n+1)x+2v(\rho+z)k-a+c}{2v}. \qquad (56)\end{aligned}$$

This version of the control equations reveals a relevant property of the game, namely that, at any time,

$$\dot{k} > 0 \quad \text{if } a-c < b(n+1)x+2v(\rho+z)k. \qquad (57)$$

In other words, this condition says that the individual R&D effort in green technologies will increase provided $a-c$ is sufficiently small. An equivalent reading is that R&D efforts will increase if the population of firms is large enough. Either way, it boils down to saying that the pace of green innovation is positively related to the intensity of competition characterising this industry.[20]

Imposing stationarity on the system $\{\dot{x}, \dot{s}, \dot{b}, \dot{k}\}$, we obtain the coordinates of the unique steady state equilibrium of the open-loop game:

$$x^* = \frac{n(a-c)z}{(n+1)\delta}; \qquad s^* = \frac{n(a-c)(\eta z-\delta)}{(n+1)\delta}; \qquad b^* = \frac{\delta}{nz}; \qquad k^* = 0. \qquad (58)$$

This steady state coincides with the monopoly equilibrium if $n=1$, and it is admissible provided that $\delta \le \eta z$. Steady state output and profits are $q^* = (a-c)/(n+1)$ and $\pi^* = (a-c)^2/(n+1)^2$, i.e., the standard Cournot-Nash profits, and $\varphi = \psi = 0$.

Now we can examine the steady state social welfare level, defined as

$$\begin{aligned}SW^* &= n\pi^* + CS^* + x^* - s^*\\ &= \frac{n(a-c)[(2(a-c+1)+n(a-c+2))\delta+2(n+1)(1-\eta)z]}{2\delta(n+1)^2}. \qquad (59)\end{aligned}$$

[20] This result has a definite Arrowvian flavour. For a summary of the debate between Schumpeter (1942) and Arrow (1962) on the relationship between market power and innovation incentives, see, e.g., Tirole (1988).

That is, in the absence of any regulation, at the steady state equilibrium welfare is the sum of industry profits, consumer surplus and the residual stock of resources, minus the amount of pollution. The effect of a change in n on welfare is described by

$$\frac{\partial SW^*}{\partial n} = \frac{(a-c)[\delta(a-c+n+1)+(n+1)(1-\eta)z]}{\delta(n+1)^3}, \tag{60}$$

$\eta \in (0, 1]$ being a sufficient condition for $\partial SW^*/\partial n > 0$ for all n. In this region (i.e., if the instantaneous regeneration rate of the natural resource is less than 100%, which, realistically, will almost always be the case), any increase in the intensity of competition generated by an increase in the population of firms is indeed beneficial. More precisely, the overall effect of a change in n on SW^* can be decomposed as follows:

$$\frac{\partial SW^*}{\partial n} = \frac{\partial(n\pi^*)}{\partial n} + \frac{\partial(CS^*)}{\partial n} + \frac{\partial x^*}{\partial n} - \frac{\partial s^*}{\partial n} \tag{61}$$

with

$$\frac{\partial(n\pi^*)}{\partial n} < 0; \qquad \frac{\partial(CS^*)}{\partial n} > 0; \qquad \frac{\partial x^*}{\partial n} > 0; \qquad \frac{\partial s^*}{\partial n} < 0, \tag{62}$$

whereby it appears that the negative effect on industry profits is more than compensated by the increase in consumer surplus, the higher volume of the natural resource and the lower level of pollution. Also, note the opposite sign of the partial derivatives of s^* and x^* w.r.t. n: a cleaner environment goes along with a higher exploitation of the natural resource in steady state, the larger the population of firms is.

If, instead, $\eta > 1$, $\partial SW^*/\partial n = 0$ in

$$n^* = \frac{(\eta-1)z-(a-c+1)\delta}{\delta-(\eta-1)z} > 1 \tag{63}$$

for all

$$\delta \in \left(\frac{2(\eta-1)z}{a-c+2}, (\eta-1)z\right). \tag{64}$$

The welfare effects of a change in the number of firms can be illustrated as in Fig. 3, drawn in the space (η, δ). Observe that

$$\eta z > (\eta-1)z \geq \frac{2(\eta-1)z}{a-c+2} \tag{65}$$

everywhere, with

$$\frac{\partial SW^*}{\partial n} > 0 \quad \forall \delta > (\eta-1)z \tag{66}$$

and

$$\frac{\partial SW^*}{\partial n} < 0 \quad \forall \delta \in \left[0, \frac{2(\eta-1)z}{a-c+2}\right), \tag{67}$$

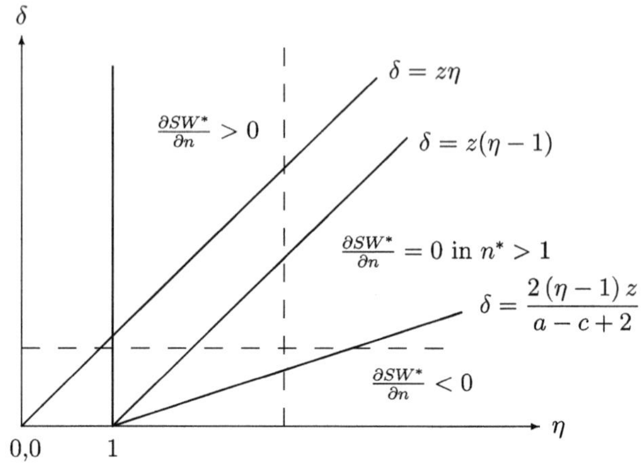

Fig. 3 Steady state welfare effects of a change in n in n in the (η, δ) space

while

$$\frac{\partial SW^*}{\partial n} = 0 \quad \text{in } n^* > 1 \tag{68}$$

for values of δ in (64). Supposing δ takes, e.g., a value like the one represented by the horizontal dashed line, the impact of industry structure on social welfare changes with η, the rate of reproduction of the natural resource. Now note, from (58), that the only steady state magnitude affected by η is s^*, in such a way that any increase in η brings about an increase in the environmental externality as it makes the resource availability constraint less stringent and therefore fosters its extraction. This, in turn, has beneficial effects on consumer surplus. For sufficiently low values of η, the positive welfare effect generated by enhancing consumer surplus exceeds the negative one associated with the parallel increase in pollution. In this range, from the standpoint of social efficiency, increasing the number of firms is desirable; accordingly, a regulator would like the market to become as competitive as possible. For sufficiently high levels of η exactly the opposite argument applies, so that monopoly is the socially efficient industry structure. In between, there exists a range where a finite number of oligopolistic firms is socially optimal. *Mutatis mutandis*, a similar argument holds if one takes an appropriate value of $\eta > 1$ as given, like the one represented by the vertical dashed line in the Figure, and evaluates what happens if δ increases. Keeping in mind that δ measures the instantaneous rate at which the environment absorbs and neutralizes pollutants, we have that, if δ is sufficiently low, monopoly maximises welfare because of the classical output restriction associated with monopoly power: here, monopoly pricing is a lesser evil. For intermediate values of δ, an oligopoly is the efficient compromise between market power and its environmental implications. Finally, if the natural process of emission recycling is sufficiently efficient, the standard pro-competitive argument prevails on environmental considerations, and therefore perfect competition maximises steady state welfare.

This discussion can be summarised in

Proposition 7 *Any $\eta \in (0, 1]$ suffices to ensure that any increase in the intensity of competition is welfare-increasing. If, instead, $\eta > 1$, the socially optimal number of firms is finite, monopoly being Pareto-efficient for all*

$$\delta \in \left[0, \frac{2(\eta - 1)z}{a - c + 2}\right).$$

An oligopoly with $n^* > 1$ firms is efficient for all

$$\delta \in \left(\frac{2(\eta - 1)z}{a - c + 2}, (\eta - 1)z\right),$$

while the Pareto-efficient structure is perfect competition for all $\delta > (\eta - 1)z$.

To ascertain the stability properties of the system, one has to inspect the following Jacobian matrix:

$$J = \begin{bmatrix} \frac{\partial \dot{x}}{\partial x} & \frac{\partial \dot{x}}{\partial s} & \frac{\partial \dot{x}}{\partial b} & \frac{\partial \dot{x}}{\partial k} \\ \frac{\partial \dot{s}}{\partial x} & \frac{\partial \dot{s}}{\partial s} & \frac{\partial \dot{s}}{\partial b} & \frac{\partial \dot{s}}{\partial k} \\ \frac{\partial \dot{b}}{\partial x} & \frac{\partial \dot{b}}{\partial s} & \frac{\partial \dot{b}}{\partial b} & \frac{\partial \dot{b}}{\partial k} \\ \frac{\partial \dot{k}}{\partial x} & \frac{\partial \dot{k}}{\partial s} & \frac{\partial \dot{k}}{\partial b} & \frac{\partial \dot{k}}{\partial k} \end{bmatrix}$$

whose eigenvalues can be easily calculated in the special case of a single firm (i.e., $n = 1$), with

$$\frac{\partial \dot{x}}{\partial x} = \eta - b; \quad \frac{\partial \dot{x}}{\partial s} = -1; \quad \frac{\partial \dot{x}}{\partial b} = -x; \quad \frac{\partial \dot{x}}{\partial k} = 0 \tag{69}$$

$$\frac{\partial \dot{s}}{\partial x} = zb - \delta; \quad \frac{\partial \dot{s}}{\partial s} = 0; \quad \frac{\partial \dot{s}}{\partial b} = zx; \quad \frac{\partial \dot{s}}{\partial k} = -1 \tag{70}$$

$$\begin{aligned} &\frac{\partial \dot{b}}{\partial x} = \frac{2[v(z(\eta + z) - v)k - bs] - (a - c)(\eta - \rho + z)}{2x^2}; \quad \frac{\partial \dot{b}}{\partial s} = \frac{b}{x}; \\ &\frac{\partial \dot{b}}{\partial b} = \frac{s + x[2(b + \eta) + \rho - z]}{x}; \quad \frac{\partial \dot{b}}{\partial k} = \frac{v[\delta - z(\eta + z)]}{x} \end{aligned} \tag{71}$$

$$\frac{\partial \dot{k}}{\partial x} = \frac{b}{v}; \quad \frac{\partial \dot{k}}{\partial s} = 0; \quad \frac{\partial \dot{k}}{\partial b} = \frac{x}{v}; \quad \frac{\partial \dot{k}}{\partial k} = \rho + z. \tag{72}$$

The resulting eigenvalues of J in $\{x^*, s^*, b^*, k^*\}$ are:

$$\varepsilon_1 = \frac{\eta + \sqrt{\eta^2 + 4\delta}}{2} > 0; \qquad \varepsilon_2 = \frac{\eta - \sqrt{\eta^2 + 4\delta}}{2} < 0$$
$$\varepsilon_3 = \frac{2\rho - \eta + \sqrt{\eta^2 + 4\delta}}{2} > 0 \tag{73}$$
$$\varepsilon_4 = -\frac{\eta - 2\rho + \sqrt{\eta^2 + 4\delta}}{2} < 0 \quad \forall \rho < \frac{\eta + \sqrt{\eta^2 + 4\delta}}{2}.$$

Accordingly, $\{x^*, s^*, b^*, k^*\}$ is a saddle point equilibrium in the monopoly case. Performing the same analysis in the general case of an oligopoly is, however, cumbersome. Yet, we can work out some numerical examples. For instance, one can fix

$$a - c = 1; \qquad n = 2; \qquad v = 1; \qquad z = 1;$$
$$\delta = \frac{1}{30}; \qquad \eta = \frac{1}{10}; \qquad \rho = \frac{1}{100} \tag{74}$$

to obtain

$$\varepsilon_1 = 0.2725; \qquad \varepsilon_2 = -0.2524; \qquad \varepsilon_3 = 0.0217; \qquad \varepsilon_4 = -0.0164, \tag{75}$$

which again reveals saddle point stability—in this case, of the duopoly equilibrium.

4 Concluding Remarks

We have modelled the dynamic interplay between firms' decisions and the resulting welfare performance of an industry involving the exploitation of a natural resource and negative environmental effects. Towards this aim, we have adopted a stepwise procedure, starting from the simplest settings to end up with a complete model including all relevant variables in a single framework. This has been done with the purpose of illustrating how some of the main properties and policy conclusions may change depending upon the degree of accuracy and completeness with which the model itself is endowed. In particular, we have focussed on the tradeoff between the opposite effects of output expansions on market price on one hand and the intensity of resource exploitation and environmental externality on the other. In this regard, a key aspect one has to bear in mind is the intensity of market competition, measured by the size of the number of active firms in the industry. The full-fledged model we have constructed indicates that, for any realistic rate of reproduction of the natural resource, any increase in the population of firms is indeed welfare-improving; conversely, monopoly is socially efficient if the emission recycling rate is very low. A related issue we have also dwelled upon is whether firms may have any incentive to invest in green technologies in the absence of taxation/subsidization. In this respect, our model suggests that such an incentive does exist, due to competitive pressure, even if firms do not explicitly internalise the effects of their activities.

Acknowledgements We would like to thank Roberto Cellini, Arsen Palestini, Tapio Palokangas (Editor), an anonymous referee and the audience at the workshop *Green Growth and Sustainable Development* (IIASA, Laxenburg, 9–10 December 2011) for useful comments and suggestions. The usual disclaimer applies.

Appendix: List of Symbols

Parameters, variables and functions appearing in text are defined as follows:

a:	reservation price
$b_i(t)$:	extraction rate of firm i
$B \equiv \sum_{i=1}^{n} b_i$:	industry extraction rate
c:	marginal production cost
CS:	consumer surplus
$\mathcal{H}_i(t)$:	Hamiltonian function of firm i
J:	Jacobian matrix
$k_i(t)$:	green R&D effort of firm i
n:	number of firms
$p(t)$:	market price
$\mathcal{P}$:	Pigouvian taxation
$q_i(t)$:	quantity of firm i
$Q(t)$:	industry output
$s(t)$:	pollution stock
SW:	social welfare
v:	marginal cost of R&D
$x(t)$:	resource stock
z:	marginal environmental damage
δ:	natural rate of emission absorption
$\gamma_i(t)$, $\varpi_i(t)$, $\zeta_i(t)$, $\varkappa_i(t)$:	co-state variables
η:	rate of reproduction of natural resources
θ:	Pigouvian tax rate
$\lambda_i(t)$, $\mu_i(t)$, $\varphi_i(t)$, $\psi_i(t)$:	co-state variables (in current value)
$\pi_i(t)$:	profits of firm i
ρ:	discount rate

References

Anderson, D. (2010). *Environmental economics and natural resource management* (3rd edn.). London: Routledge.

Arrow, K. J. (1962). Economic welfare and the allocation of resources for invention. In R. Nelson (Ed.), *The rate and direction of industrial activity*. Princeton: Princeton University Press.

Arrow, K. J. (1968). Applications of control theory to economic growth. In G. B. Dantzig & A. F. Veinott (Eds.), *Mathematics of the decision sciences part 2*. Providence: Am. Math. Soc.

Benchekroun, H., & Long, N. V. (1998). Efficiency inducing taxation for polluting oligopolists. *Journal of Public Economics*, *70*, 325–342.
Benchekroun, H., & Long, N. V. (2002). On the multiplicity of efficiency-inducing tax rules. *Economics Letters*, *76*, 331–336.
Chiou, J.-R., & Hu, J.-L. (2001). Environmental research joint ventures under emission taxes. *Environmental and Resource Economics*, *20*, 129–146.
Clark, C. W., & Munro, G. R. (1975). The economics of fishing and modern capital theory. *Journal of Environmental Economics and Management*, *2*, 92–106.
Copeland, B. R., & Taylor, S. M. (2009). Trade, tragedy, and the commons. *American Economic Review*, *99*, 725–749.
Cornes, R., & Sandler, T. (1983). On commons and tragedies. *American Economic Review*, *73*, 787–792.
Cornes, R., Mason, C. F., & Sandler, T. (1986). The commons and the optimal number of firms. *Quarterly Journal of Economics*, *101*, 641–646.
Dockner, E. J., Jørgensen, S., Long, N. V., & Sorger, G. (2000). *Differential games in economics and management science*. Cambridge: Cambridge University Press.
Downing, P. B., & White, L. J. (1986). Innovation in pollution control. *Journal of Environmental Economics and Management*, *8*, 225–271.
Gordon, H. S. (1954). The economic theory of a common-property resource: the fishery. *Journal of Political Economy*, *62*, 124–142.
Hardin, G. (1968). The tragedy of the commons. *Science*, *162*, 1243–1248.
Karp, L., & Livernois, J. (1994). Using automatic tax changes to control pollution emissions. *Journal of Environmental Economics and Management*, *27*, 38–48.
Levhari, D., & Mirman, L. (1980). The great fish war: an example using a dynamic Cournot-Nash solution. *Bell Journal of Economics*, *11*, 322–334.
Long, N. V. (2010). *A survey of dynamic games in economics*. Singapore: World Scientific.
Mangasarian, O. L. (1966). Sufficient conditions for the optimal control of nonlinear systems. *SIAM Journal on Control*, *4*, 139–152.
Mason, C. F., & Polasky, S. (1997). The optimal number of firms in the commons: a dynamic approach. *Canadian Journal of Economics*, *30*, 1143–1160.
Milliman, S. R., & Prince, R. (1989). Firm incentives to promote technological change in pollution control. *Journal of Environmental Economics and Management*, *17*, 247–265.
Pearce, D. W., & Turner, R. K. (1989). *The economics of natural resources and the environment*. Hemel Hempstead: Harvester Wheatsheaf.
Poyago-Theotoky, J. A. (2007). The organization of R&D and environmental policy. *Journal of Economic Behavior and Organization*, *62*, 63–75.
Schumpeter, J. A. (1942). *Capitalism, socialism and democracy*. New York: Harper & Row.
Stern, N. (2007). *The economics of climate change: the Stern Review*. Cambridge: Cambridge University Press.
Stern, N. (2009). *A blueprint for a safer planet: how to manage climate change and create a new era of progress and prosperity*. London: Bodley Head.
Tirole, J. (1988). *The theory of industrial organization*. Cambridge: MIT Press.
Tisdell, C. (2009). *Resource and environmental economics. modern issues and applications*. Singapore: World Scientific.
Weitzman, M. L. (2007). A review of the Stern Review on the economics of climate change. *Journal of Economic Literature*, *45*, 703–724.

The Incentive to Invest in Environmental-Friendly Technologies: Dynamics Makes a Difference

Davide Dragone, Luca Lambertini, and Arsen Palestini

1 Introduction

The enormous amount of data being assembled by the IPCC (Intergovernmental Panel on Climate Change) on the anthropic responsibility in generating (or at least increasing) global warming, and the debate on how to cope with it along the guidelines of the Kyoto Protocol and its follow-ups, are clearly identifying the control of polluting emissions damaging the environment as one of the hottest scientific issues of our times. As such, it is receiving an increasing amount of attention in the current literature in the field of environmental economics, with particular attention to the general equilibrium implications of environmental aspects on trade and growth.[1]

Most of the existing contributions adopting a partial equilibrium approach investigate the design of optimal Pigouvian taxation aimed at inducing firms to reduce damaging emissions, both in monopoly and oligopoly settings.[2] A related stream of

[1]On the optimality of free trade with environmental externalities, see Copeland and Taylor (1994, 2004) and Antweiler et al. (2001). As to the role of environmental issues in growth theory, see Grossman and Krueger (1995), Bovenberg and de Mooij (1997), Bartz and Kelly (2008), Itaya (2008) and Dragone et al. (2010), inter alia.

[2]See Karp and Livernois (1994) and Benchekroun and Long (1998, 2002), inter alia.

D. Dragone (✉) · L. Lambertini
Department of Economics, University of Bologna, Strada Maggiore 45, Bologna 40125, Italy
e-mail: davide.dragone@unibo.it

L. Lambertini
e-mail: luca.lambertini@unibo.it

L. Lambertini
ENCORE, University of Amsterdam, Roeterstraat 11, Amsterdam WB1018, The Netherlands

A. Palestini
MEMOTEF, Sapienza University of Rome, Via del Castro Laurenziano 9, Rome 00161, Italy
e-mail: arsen.palestini@uniroma1.it

J. Crespo Cuaresma et al. (eds.), *Green Growth and Sustainable Development*, Dynamic Modeling and Econometrics in Economics and Finance 14,
DOI 10.1007/978-3-642-34354-4_8,

literature examines the incentive for firms to carry out R&D activities in order to introduce environmental-friendly technologies. This of course is very closely related to trade and growth. In particular, a sustainable development will require cheaper and cleaner energy sources and productive technologies than are currently available, to be attained through innovation.[3] In static setups, this requires the introduction of some form of taxation/subsidy by the policy maker, in order to induce firms to take into account the presence of the externality, that they would clearly neglect otherwise.[4] A third line of research investigates the optimal design of minimum quality standards and/or profit taxation in vertically differentiated industries affected by environmental externalities.[5]

In the present paper, we take a differential game approach to the investigation of environmentally-oriented R&D efforts in a dynamic Cournot oligopoly model where (at least in the first version of the game) there may not be any tax or subsidy linked to the external effect, in order to show that the main message emerging from the corresponding static version of the same game falls short of telling the whole story of the issue at hand. In particular, we describe a scenario where the stock of pollution increases in proportion to industry output, and each firm may invest in R&D in order to diminish its individual contribution to the emission of pollutants.

Our first result consists in showing that unregulated firms may indeed fully neglect the environmental effects of their productive activity and replicate the static Cournot-Nash equilibrium forever, without putting any effort whatsoever in R&D activities for cleaner technologies at any point in time. However, we also show that the alternative may in fact be more attractive, if R&D efforts go along with an output contraction closely resembling cartel behaviour, although the setup remains fully non cooperative. That is, we identify a path along which, by taking explicitly into account the externality, firms performs environmental R&D investments not because of some altruistic or environmental concern but for pure profit-seeking reasons.[6]

The game among unregulated firms yields multiple steady state equilibria, all of them (except of course the quasi-static solution replicating the Cournot outcome forever) being characterised by positive R&D efforts at all times, except possibly doomsday. In summary, the appraisal of our analysis of private incentives can be outlined as follows. First, the static game captures the main feature of one of the steady states we identify, but cannot grasp the essence of what happens along the optimal path to this long run equilibrium. Secondly, the remaining two equilibria, both emerging whenever the stock of polluting emissions vanishes, are linked by

[3]See, e.g., Klemperer (2007).

[4]To this regard, see Downing and White (1986), Milliman and Prince (1989), Damania (1996), Scott (1996), Chiou and Hu (2001), Mohr (2002), Hart (2004), Greaker (2006) and Poyago-Theotoky (2007), inter alia.

[5]See Lutz et al. (2000), Amacher et al. (2004), Lombardini-Riipinen (2005), André et al. (2009) and Bottega and De Freitas (2009), inter alia.

[6]In a similar setting, Benchekroun and Chaudury (2011) show that imposing a Markovian tax on emissions may bring about a stable cartel, while this does not happen with a uniform tax.

saddle point trajectories which exit the least preferable point to enter the most desirable one, as far as profit, consumer surplus and social welfare are concerned. This is a desirable property, entirely driven by profit incentives, which in the present case are not in conflict with social preferences.

Then, we examine two modified versions of our setup: in the first one, a social planner concentrates the production of the good in a unique plant, whereby the activity of R&D takes place in N different structures (due to the decreasing returns to scale characterising the R&D technology). In this case, five steady state points exist, one of which replicates the perfectly competitive allocation that would emerge under social planning in the corresponding static version of the model. Yet, a relevant feature of this equilibrium is that the planner would be able to reach it only in the very specific (and totally unrealistic) case where the production of the final good were not polluting the environment at all.

The second extension takes into account the possibility of regulating profit-seeking firms via the introduction of a Pigouvian tax associated to the environmental externality. In this case, we show that the tax can be designed so as to induce the industry to yield the first best level of social welfare that is unattainable under planning, although of course the associated surplus distribution is not the same as it would be at the first best.

The remainder of the paper is structured as follows. Section 2 briefly outlines the static version of the game. The setup of the dynamic problem and the related trajectory analysis are laid out in Sect. 3, where we also compare the profit and welfare performance of the industry in correspondence of the multiple steady state equilibria. In Sect. 4 we examine the behaviour of the model under social planning. In Sect. 5 we illustrate the effects of Pigouvian taxation on the equilibrium behaviour of profit-seeking firms as well as the related welfare levels and provide an interpretation for such a tax. Section 6 contains our concluding remarks.

2 A Summary of the Static Problem

As a preliminary step, we revisit the static Cournot game in order to highlight the lack of R&D incentives to decrease the amount of polluting emissions characterising firms. The market is supplied by N single-product homogeneous-good firms. The market inverse demand function is $p = a - Q$, with $Q = \sum_{i=1}^{N} q_i$, q_i being firm i's output. Technology is the same for all firms alike, and it is summarised by the cost function $C = cq_i$. Supplying the final good entails a negative environmental externality $S = \sum_{i=1}^{N} b_i q_i$, where $b_i = \overline{b} - k_i \geq 0$; $\overline{b}$ measures the marginal contribution of each firm to the stock of pollutants; k_i is the R&D effort of firm i to decrease its individual amount of pollution,[7] and it involves a convex cost $\Gamma_i = rk_i^2$, $r > 0$.

[7] Here we assume firm-specific externalities and R&D activities, as it appears to be reasonable in examining investments in environmental-friendly technologies. Hence, we rule out the possibility of spillovers in R&D.

Consequently, firm i's instantaneous profits are $\pi_i = (p - c)q_i - \Gamma_i$. This game has a two-stage structure: in the first stage, firms non-cooperatively and simultaneously set their respective R&D efforts; in the second, they compete à la Cournot-Nash. The solution concept is subgame perfection by backward induction.

The optimal individual output in the second stage is $q^* = (a - c)/(N + 1)$, whereby the profit function at the first stage reads as $\pi_i = (q^*)^2 - rk_i^2$. This clearly entails that $\partial\pi_i/\partial k_i < 0$, and therefore the optimal R&D investment is nil, yielding the static Cournot-Nash profits $\pi^{CN} = (q^*)^2$. On this basis, one has to introduce some form of environmental taxation, no matter whether it is firm-specific or not, to induce firms to take into account the presence of the externality and indeed carry out some R&D efforts to reduce it. As we shall see in the following sections, this is not necessarily the case if one adopts a properly dynamic approach to this issue.

3 The Dynamic Setup

As in the static model, consider a Cournot oligopoly with N single-product homogeneous-good firms interacting over continuous time $t \in [0, \infty)$. At any time t, the demand function is $p(t) = a - Q(t)$, with $Q(t) = \sum_{i=1}^{N} q_i(t)$, $q_i(t)$ being the instantaneous individual output of firm i. All firms use the same productive technology, described by the cost function $C_i(t) = cq_i(t)$. The production of the final output involves a negative environmental externality $S(t)$, evolving according to the following dynamics:

$$\dot{S}(t) = \frac{dS}{dt} = \sum_{i=1}^{N} b_i(t)q_i(t) - \delta S(t), \tag{1}$$

where $\delta > 0$ is a constant decay rate and $S(0) = S_0 > 0$ is the initial condition. The coefficient $b_i(t) \geq 0$, with $b_i(0) = b_{i0} \geq 0$, measures the marginal contribution to the stock of pollution that the production of firm i entails. Depending on the R&D effort $k_i(t)$ of i, it evolves over time according to the following equation:

$$\dot{b}_i(t) = b_i(t)\big[\eta - k_i(t)\big], \quad \eta > 0. \tag{2}$$

That is, until k_i is smaller than the threshold value η, b_i is increasing. As in the static game, the instantaneous cost associated with the R&D activity is $\Gamma_i(t) = r(k_i(t))^2$, with $r > 0$. Hence, firm i's instantaneous profits are $\pi_i(t) = [p(t) - c]q_i(t) - \Gamma_i(t)$, and each firm i has to set $q_i(t)$ and $k_i(t)$ so as to maximise

$$\Pi_i = \int_0^{\infty} \big\{\big[p(t) - c\big]q_i(t) - \Gamma_i(t)\big\}e^{-\rho t}dt, \tag{3}$$

under the state equations (1) and (2) and the initial conditions. Parameter $\rho > 0$ is a constant discount rate common to all firms.

3.1 Equilibrium Analysis

From now on, we will omit the time argument for simplicity, whenever possible. The solution concept is the open-loop Nash equilibrium.[8] The current-value Hamiltonian of firm i is:

$$\begin{aligned}\mathcal{H}_i(\cdot) &= [p-c]q_i - \Gamma_i + \lambda_i \dot{S} + \mu_{ii}\dot{b}_i + \sum_{j\neq i}\mu_{ij}\dot{b}_j \\ &= (\sigma - Q)q_i - rk_i^2 + \lambda_i \dot{S} + \mu_{ii}\dot{b}_i + \sum_{j\neq i}\mu_{ij}\dot{b}_j,\end{aligned} \tag{4}$$

where $\sigma \equiv a - c > 0$ denotes the market dimension.

The necessary conditions (FOCs) are:

$$\frac{\partial \mathcal{H}_i}{\partial q_i} = \sigma - 2q_i - Q_{-i} + \lambda_i b_i = 0, \tag{5}$$

where $Q_{-i} \equiv \sum_{j\neq i} q_j$, and

$$\frac{\partial \mathcal{H}_i}{\partial k_i} = -2rk_i - \mu_{ii} b_i = 0. \tag{6}$$

Note that no $\mu_{ij}(t)$ appears in the FOCs, thus it does not affect the model's equilibrium structure. The adjoint equations read as follows:

$$\dot{\lambda}_i = (\rho + \delta)\lambda_i \tag{7}$$

$$\dot{\mu}_{ii} = [\rho - \eta + k_i]\mu_{ii} - \lambda_i q_i \tag{8}$$

$$\dot{\mu}_{ij}(t) = \left[\rho - \eta + k_j(t)\right]\mu_{ij}(t) - \lambda_i(t) q_j(t). \tag{9}$$

From (5) and (6) one obtains, respectively:

$$\lambda_i = -\frac{\sigma - 2q_i - Q_{-i}}{b_i}, \tag{10}$$

$$\mu_{ii} = -\frac{2rk_i}{b_i}. \tag{11}$$

The associated (necessary) transversality conditions are:

[8] Note that, since (2) contains a product between a control and a state, the model is not a linear-quadratic one, and therefore there exists no obvious candidate for the optimal value function that one should adopt to solve the feedback game.

$$\begin{aligned}
&\lim_{t\to\infty} e^{-\rho t}\lambda_i(t)S(t) = 0;\\
&\lim_{t\to\infty} e^{-\rho t}\mu_{ii}(t)b_i(t) = 0;\\
&\lim_{t\to\infty} e^{-\rho t}\mu_{ij}(t)b_j(t) = 0.
\end{aligned} \tag{12}$$

Before carrying out the equilibrium analysis, it is worth dwelling upon the interpretation of the above necessary conditions. First of all, note that (7) admits the solution $\lambda_i(t) = 0$ at all times, which in turn allows $\mu_{ii}(t) = 0$ to be a solution to (8). In such a case, the dynamic model would immediately reproduce the very same outcome of the static game, with no investments at all at any time and the static Cournot-Nash equilibrium replicated at all t:

Proposition 1 *Adjoint equations admit the solution $\lambda_i(t) = \mu_{ii}(t) = 0$ at all $t \in [0, \infty)$. This entails $q_i(t) = \sigma/(N+1)$ and $k_i(t) = 0$ for all $i = 1, 2, 3, \ldots, N$ at all $t \in [0, \infty)$.*

However, if the R&D control is always nil and the output control is always equal to the static Cournot-Nash solution, the level of pollution would explode to plus infinity unless $b_i(0) = b_{i0} = 0$, i.e., unless the polluting features of productive technology are not an issue because technology itself is already clean at the very outset (which of course makes the entire story a trivial one).

Additionally, adjoint equations (7)–(8) also admit non-nil solutions which, by definition, do not appear in the static version of the game. This has some interesting implications as to the firms' incentive to invest in environmental-friendly technologies. To shed light on this aspect, we may propose the following observations.

Equation (5) produces firm i's instantaneous best reply:

$$q_i^*(Q_{-i}) = \frac{\sigma - Q_{-i} + \lambda_i b_i}{2} \tag{13}$$

This yields firm i's optimal output for any given vector of the rivals' outputs, irrespective of the value of the costate vector $\lambda_{-i}(t)$. In particular, note that (13) does not convey any information as to $\lambda_{-i}(t)$, but simply instructs firm i as to the eventual profitability of a strategy involving a non-nil shadow value $\lambda_i(t)$. To assess this perspective, we proceed as follows.

The best reply $q_i^*(Q_{-i})$ shifts inwards (resp. outwards) w.r.t. its static counterpart for all $\lambda_i < 0$ (resp., $\lambda_i > 0$). Equivalently, (10) takes a negative value for all $Q < N\sigma/(N+1)$, i.e., whenever the industry output is lower than its static Cournot-Nash level (and conversely). Now, if $\lambda_i < 0$, the inward shift of best reply functions entails an output contraction that, nonetheless, is driven by a fully non cooperative behaviour. Also, note that (8) yields $\mu_{ii} < 0$ for all $k_i > 0$. The fact that adjoint variables are negative indicates that firm i attaches a negative shadow value to its marginal contribution to the increase in the pollution stock. Yet, the output contraction opens the possibility that the firm increases its profits instant by instant, even if a costly R&D project for a greener technology is undertaken.[9] That is, the incentive to adopt the investment strategy associated with $\lambda_i(t) < 0$ is highlighted by the flow of instantaneous gains exemplified by:

$$\pi^{CN}(k=0) \equiv \frac{\sigma^2}{(N+1)^2} < (\widehat{p}-c)\widehat{q} - rk^2 \equiv \widehat{\pi} \quad (k>0) \tag{14}$$

for non-empty sets of values of $k > 0$ and $\widehat{q} \in (0, \frac{\sigma}{N+1})$. During the game, firm i may smooth the R&D investment not because she has developed any environmentally-oriented conscience of her own, but rather in order to be able to keep the output at a quasi-collusive level forever. In other words, from the firms' viewpoint, the R&D cost Γ_i is the fee to be paid to build up a path replicating that of a cartel in quantities, without actually taking any implicitly collusive attitude that would constitute a target for the antitrust authority.[10] Conversely, from consumers' viewpoint, a higher market price is what they have to pay in return for a cleaner environment.

Having said that, we may proceed to the characterisation of the equilibrium behaviour. One can impose symmetry across quantities, costate variables and states:

$$q_i = q_j = q, \qquad \lambda_i = \lambda_j = \lambda, \tag{15}$$

$$\mu_{ii} = \mu_{jj} = \mu, \qquad b_i = b_j = b. \tag{16}$$

From the FOCs (5) and (6) one also obtains the control equations:

$$\dot{q} = \frac{\lambda\dot{b} + \dot{\lambda}b}{N+1}, \qquad \dot{k} = -\frac{\mu\dot{b} + \dot{\mu}b}{2r} \tag{17}$$

which can be rewritten, using (7)–(8) and (10)–(11), leading to the following state-control dynamical system:

$$\dot{S} = Nbq - \delta S \tag{18}$$

$$\dot{b} = b(\eta - k) \tag{19}$$

$$\dot{q} = \frac{[(N+1)q - \sigma][\rho + \delta + \eta - k]}{N+1} \tag{20}$$

$$\dot{k} = \rho k - \frac{q[\sigma - (N+1)q]}{2r} \tag{21}$$

Although the equations (18)–(19) and (20)–(21) are not decoupled, we can stress that, given any solution curve $(q^*(t), k^*(t))$ of equations (20)–(21), we can obtain

[9] Using a repeated game with infinite Nash reversion, Damania (1996) finds that firms may not be willing to buy pollution-abating technologies if the associated exogenous cost is too high.

[10] Moreover, this eliminates any issue concerning the possibility of unilateral deviations, as it is the outcome of a fully noncooperative behaviour.

the state trajectories by applying the methods of separation of variables and Lagrange's variation of constants to (18)–(19):

$$b^*(t) = b_0 e^{\eta t - \int_0^t k^*(s)ds}, \tag{22}$$

$$S^*(t) = \left(S_0 + b_0 \int_0^t \left(e^{(\eta+\delta)s - \int_0^s k^*(\tau)d\tau}\right) q^*(s)ds \right) e^{-\delta t}. \tag{23}$$

In particular, note that if the costate variables are not identically zero, then the transversality condition concerning $S(t)$ becomes:

$$\lim_{t\to\infty} e^{-\rho t} \lambda_{i0} e^{(\rho+\delta)t} \left(S_0 + b_0 \int_0^t \left(e^{(\eta+\delta)s - \int_0^s k^*(\tau)d\tau}\right) q^*(s)ds \right) e^{-\delta t}$$
$$= \lim_{t\to\infty} \lambda_{i0} \left(S_0 + b_0 \int_0^t \left(e^{(\eta+\delta)s - \int_0^s k^*(\tau)d\tau}\right) q^*(s)ds \right) = 0$$

if and only if $S_0 = -b_0 \int_0^\infty (e^{(\eta+\delta)s - \int_0^s k^*(\tau)d\tau}) q^*(s)ds$, meaning that this is the relation between the initial conditions of the state variables leading to a non-trivial equilibrium structure. That is, when such relation holds, the optimal trajectories will approach the steady states described in the following Proposition.

Expressions (22) and (23) imply that both $b_i^*(t)$ and $S^*(t) > 0$ are non negative at all times (except, possibly, doomsday in which they are nil) Before inspecting the stationary points of the above dynamic system, it is worth observing that, using the time elimination method, we can write the derivatives ratio:

$$\frac{\dot{q}}{\dot{k}} = \frac{2r[(N+1)q - \sigma][\rho + \delta + \eta - k]}{(N+1)[2r\rho k - q(\sigma - (N+1)q)]} = \frac{dq}{dk} \tag{24}$$

indicating the slope of the open-loop Nash trajectory in the control plane.

The sign of (24) is evaluated in

Remark 2 Take $q \in (0, \frac{\sigma}{N+1})$. Then, $dq/dk < 0$ for all

$$k \in \left(\min\left\{ \rho + \delta + \eta, \frac{q(t)[\sigma - (N+1)q]}{2r\rho} \right\}, \max\left\{ \rho + \delta + \eta, \frac{q[\sigma - (N+1)q]}{2r\rho} \right\} \right).$$

That is to say, for any individual output level lower than the Cournot-Nash output, there exists an admissible range of values for k wherein the two controls are substitutes at a generic point in time, during the game. In such a case, any output contraction with respect to the Cournot-Nash static equilibrium drives some R&D effort for cleaner technologies.

Steady state equilibria are described by the following:

Proposition 3 *The stationary points of the system are*:

$$P_A = (S_A, b_A, q_A, k_A) = \left(0, 0, \frac{\sigma}{N+1}, 0\right),$$
$$P_B = (S_B, b_B, q_B, k_B) = (0, 0, q_B, \delta + \rho + \eta),$$
$$P_C = (S_C, b_C, q_C, k_C) = (0, 0, q_C, \delta + \rho + \eta),$$

where

$$q_B = \frac{\sigma - \sqrt{\sigma^2 - 8r(N+1)(\rho+\delta+\eta)\rho}}{2(N+1)},$$
$$q_C = \frac{\sigma + \sqrt{\sigma^2 - 8r(N+1)(\rho+\delta+\eta)\rho}}{2(N+1)}.$$

Proof Imposing the stationarity condition $\dot{k} = 0$ yields

$$k(q) = \frac{q[\sigma - (N+1)q]}{2r\rho} \tag{25}$$

which can be plugged into $\dot{q} = 0$ to obtain the following solutions:

$$q_A = \frac{\sigma}{N+1}; \qquad q_{B,C} = \frac{\sigma \pm \sqrt{\sigma^2 - 8r(N+1)(\rho+\delta+\eta)\rho}}{2(N+1)} \tag{26}$$

with $q_{B,C} \in \mathbb{R}_+$ for $\sigma > \sqrt{8r(N+1)(\rho+\delta+\eta)\rho}$. By substituting in (25) we have that $k_{B,C} = \delta + \rho + \eta$.

In correspondence of the Cournot-Nash optimal quantity q_A, we have $k_A = 0$, $S_A = 0$, $b_A = 0$. □

The following results show the dynamic behaviour of the optimal solutions:

Proposition 4 P_A, P_B *and* P_C *are saddle points of the system.*

Proof The Jacobian matrix of the state-control system reads as:

$$J = \begin{pmatrix} -\delta & Nq & Nb & 0 \\ 0 & \eta - k & 0 & -b \\ 0 & 0 & \rho + \delta + \eta - k & -q + \dfrac{\sigma}{N+1} \\ 0 & 0 & \dfrac{1}{2r}[2(N+1)q - \sigma] & \rho \end{pmatrix}. \tag{27}$$

$J(P_A)$ has the eigenvalues $\lambda_1 = -\delta < 0$, $\lambda_2 = \eta > 0$, $\lambda_3 = \rho + \delta + \eta > 0$ and $\lambda_4 = \rho > 0$, subsequently P_A is a saddle point.

The analysis of the remaining two equilibria is slightly more difficult: both $J(P_B)$ and $J(P_C)$ admit the negative eigenvalues $\lambda_1 = -\delta < 0$ and $\lambda_2 = -\rho - \delta < 0$, so the stability properties of those two points depend on the roots of the characteristic polynomials of the submatrices, for $j = B, C$:

$$\begin{pmatrix} \rho + \delta + \eta - k_j & -q_j + \dfrac{\sigma}{N+1} \\ \dfrac{1}{2r}\left[2(N+1)q_j - \sigma\right] & \rho \end{pmatrix}, \tag{28}$$

i.e.

$$p_j(\lambda) = \lambda^2 - \rho\lambda - \frac{1}{2r}\left(-q_j + \frac{\sigma}{N+1}\right)\left[2(N+1)q_j - \sigma\right]. \tag{29}$$

If $j = B$, the two remaining eigenvalues are complex with real part $\rho/2 > 0$, whereas if $j = C$, they are real and at least one of them is positive, hence P_B and P_C are saddle points too. □

In the above Proposition, we have used the term saddle point with reference to the presence of eigenvalues with different signs. However, as is well known, a saddle point can be reached starting from initial states that can be subject to more or less stringent conditions. In particular, this requires delving into the details of P_A:

Remark 5 The steady state P_A is degenerate, as it can be reached only along an equilibrium trajectory which solves (18)–(19) for $b_0 = 0$ and for any $S_0 > 0$, i.e., it is completely contained in the half-line determined by the intersection of the hyperspaces $b = 0$, $q = \frac{\sigma}{N+1}$, $k = 0$, with the stock of pollution asymptotically decreasing to 0.

That is, the equilibrium reproducing the Cournot-Nash outcome can be attained iff the technology is already fully environmental-friendly from the outset, which makes this case quite peculiar and somewhat uninteresting. Or, put it in other terms, the requirement on b_0 indicates that the prediction of the static game is far from convincing. Completely different considerations apply to the remaining two steady states, that are attainable for $b_0 > 0$.

Proposition 6 *In the half-space $k > \eta$, along each equilibrium trajectory of the system close to P_B and P_C the state variables S and b are monotonically decreasing to* 0.

Proof The stationary points P_B and P_C belong to the half-space $k > \eta$. The eigenvectors of $J(P_B)$ and $J(P_C)$ imply that the stable subspaces $E_s(P_B)$ and $E_s(P_C)$ are spanned by the vectors of the canonical basis of $\mathbb{R}^4$: $(1, 0, 0, 0)$ and $(0, 1, 0, 0)$, that is the trajectories on the respective stable manifolds are heading towards the equilibrium coordinates $S = 0$, $b = 0$. □

Fig. 1 On the control plane, the saddle point trajectories either leave the Cournot-Nash equilibrium P_A or spiral around P_B. The feasibility of P_C is ensured

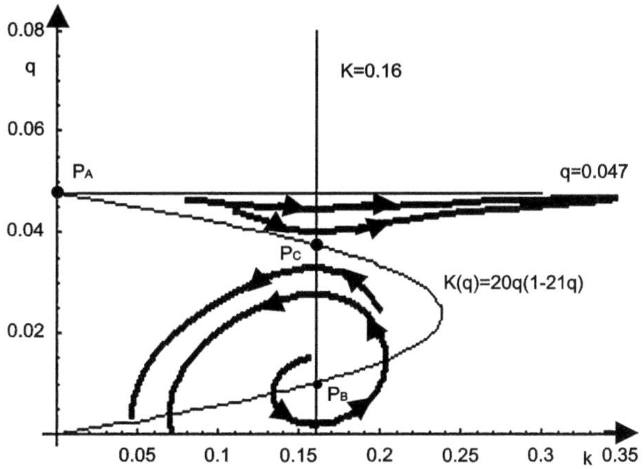

The economic meaning of the previous results is clear: in correspondence of the two points P_B and P_C the stock of pollution tends to diminish and finally disappears.

From the standpoint of the dynamical behaviour of the system, in the above-mentioned half-space the Nash trajectories approach P_B in the control plane, spiral around it and then head towards P_C, which is a saddle point in the sense that there exists a phase curve contained in the control plane which enters P_C. As we will see in next subsection, this is good news because in that point higher levels of profit and social welfare can be reached with respect to P_B.

The figure we are showing is sketched with the help of Mathematic@ 5.0, after suitably setting the relevant parameters:

$$N = 20, \qquad \sigma = 1, \qquad \rho = 3 \cdot 10^{-2},$$
$$\eta = 10^{-2}, \qquad r = 10^{-2}, \qquad \delta = 10^{-2}.$$

In the plot we can visualize the sketches of some equilibrium trajectories on the (k, q) control plane, with the same parameter values as in Fig. 1. On such a plane, coherently with the eigenvalues of (28), $P_B = (0.16, 0.08)$ is clearly an unstable focus, whereas $P_C = (0.16, 0.38)$ is a saddle point.

Moreover, the optimal R&D effort of the representative firm is positive at any time t during the game. Or, put it the other way around, any non-zero value of the co-state variable attached to the dynamics of the individual firm's contribution to the increase of the pollution stock ensures that the firm itself has indeed an incentive to invest in R&D activities for pollution abatement all along the game.

3.2 *Profit and Welfare Assessment*

In this section we compare the optimal quantities, the level of profits and of social welfare associated to the three steady states.

Proposition 7 *For every admissible* σ, N, r, ρ, δ, η, *we have* $q_A > q_C > q_B$.

In steady state, the profit levels are the following:

$$\pi(P_A) = \frac{\sigma^2}{(N+1)^2}, \tag{30}$$

$$\pi(P_B) = \sigma q_B - N q_B^2 - r(\rho + \delta + \eta)^2, \tag{31}$$

$$\pi(P_C) = \sigma q_C - N q_C^2 - r(\rho + \delta + \eta)^2. \tag{32}$$

On the basis of (30)–(32), we can state:

Proposition 8 *The profits* $\pi(P_B)$ *and* $\pi(P_C)$ *are positive if either of the following holds*:

1. $\rho \geq \delta + \eta$;
2. $\rho < \delta + \eta$ *and*

$$2\sqrt{2(N+1)(\delta + \eta - \rho)} < \sigma < \left[(N+1)(\delta + \eta) + (1-N)\rho\right]\sqrt{\frac{r(\delta + \eta + \rho)}{\delta + \eta - \rho}}.$$

Assuming that the parameters are such that profits are indeed non negative, we can make a comparison to assess the relative desirability of the three outcomes:

Proposition 9 *The following inequalities hold*:

1. $\pi(P_C) > \pi(P_B)$ *irrespective of parameter values*;
2. $\pi(P_A) > \pi(P_C)$ *if* $\rho \in [0, \delta + \eta)$.

The intuition behind the above result is that P_A is characterised by a larger output level (which, per se, would be detrimental for profits) but the corresponding R&D effort is nil (which in turn is good news for profits), while the remaining two steady states are characterised by lower output levels in combination with positive R&D efforts. In particular, it is noteworthy observing that the Cournot-Nash solution may be worse than the steady state P_C where the firm indeed invests in R&D, despite the fact that pollution does not affect its profits.

Now we turn to consumer surplus $CS(P_i)$, $i = A, B, C$, in the three equilibria. Note that, in principle, the definition of consumer surplus would be

$$CS(P_i) = \frac{Q_i^2}{2} - S_i, \tag{33}$$

where Q_i is the sum of outputs and S_i is the pollution stock at the i-th steady state. Since $S = 0$ always in steady state, we can summarise the resulting ranking as follows:

Proposition 10 *Over the entire admissible range of parameters, we have* $CS(P_A) > CS(P_C) > CS(P_B)$.

Finally, we are going to evaluate social welfare. Assuming symmetry throughout all the oligopolists, the social welfare function $SW(P_i)$ evaluated at the steady state P_i reads as:

$$\begin{aligned} SW(P_i) &= N\pi(P_i) + CS(P_i) \\ &= N\left[(\sigma - Nq_i)q_i - rk_i^2\right] + \frac{N^2 q_i^2}{2} - S_i \\ &= N\left(\sigma q_i - rk_i^2\right) - \frac{N^2 q_i^2}{2} - S_i, \end{aligned} \tag{34}$$

for $i = A, B, C$, to obtain:

Proposition 11 *Over the entire admissible range of parameters, we have* $SW(P_A) > SW(P_C) > SW(P_B)$.

Propositions 7–11 also entail:

Corollary 12 *Any* $\rho \in [0, \delta + \eta)$ *suffices to ensure that private and social preferences over the spectrum of steady state equilibria are reciprocally aligned.*

This essentially relies upon the fact that the industry R&D effort in P_A is nil. Note however that, as we have outlined above, P_A is indeed degenerate.

4 Social Planning

We assume that the benevolent planner uses a single plant for the production of the consumption good (in view of the constant returns to scale characterising the related technology), while keeping N R&D labs, as this activity features decreasing returns. Hence, the list of variables reduces to $N + 1$ controls and two states, namely, S and b. The Hamiltonian of the planner is:[11]

$$\mathcal{H}_{SP}(\cdot) = \left\{(\sigma - q)q + \frac{q^2}{2} - S - Nrk^2 + \lambda(bq - \delta S) + \mu b(\eta - k)\right\} \tag{35}$$

[11] We attribute to the planner the same time discounting that we have used to measure firms's time preferences in the previous section. One might, however, suppose that the planner's discount rate be significantly lower than firms (possibly even nil), in order to give an appropriate weight to the welfare of future generations. For a thorough appraisal of this issue, see the *Stern Review* (Stern 2007) as well as Dasgupta (2007), Nordhaus (2007) and Weitzman (2007).

where subscript SP stands for *social planning*. The necessary conditions are:

$$\frac{\partial \mathcal{H}_{SP}}{\partial q} = \sigma - q + \lambda b = 0; \tag{36}$$

$$\frac{\partial \mathcal{H}_{SP}}{\partial k} = -N(\mu b + 2rk) = 0; \tag{37}$$

$$-\frac{\partial \mathcal{H}_{SP}}{\partial S} = \dot{\lambda} - \rho\lambda \quad \Leftrightarrow \quad \dot{\lambda} = (\rho + \delta)\lambda + 1; \tag{38}$$

$$-\frac{\partial \mathcal{H}_{SP}}{\partial b} = \dot{\mu} - \rho\mu \quad \Leftrightarrow \quad \dot{\mu} = (\rho - \eta + Nk)\mu - \lambda q. \tag{39}$$

With respect to the case of competition, observe that, under social planning, in steady state it cannot be that $\lambda = \mu = 0$. By manipulating the above conditions, we obtain the following state-control system:

$$\dot{S} = bq - \delta S \tag{40}$$

$$\dot{b} = b(\eta - Nk) \tag{41}$$

$$\dot{q} = b + (q - \sigma)(\rho + \delta + \eta - Nk) \tag{42}$$

$$\dot{k} = \rho k - \frac{q(\sigma - q)}{2r} \tag{43}$$

Unlike the oligopoly game we have investigated above, the planner's problem yields five steady state points:

Proposition 13 *The stationary points of the system are*:

$$P_{SP1} = (S_{SP1}, b_{SP1}, q_{SP1}, k_{SP1}) = (0, 0, \sigma, 0),$$

$$P_{SP2} = (S_{SP2}, b_{SP2}, q_{SP2}, k_{SP2}) = \left(0, 0, q_{SP2}, \frac{\delta + \rho + \eta}{N}\right),$$

$$P_{SP3} = (S_{SP3}, b_{SP3}, q_{SP3}, k_{SP3}) = \left(0, 0, q_{SP3}, \frac{\delta + \rho + \eta}{N}\right),$$

$$P_{SP4} = (S_{SP4}, b_{SP4}, q_{SP4}, k_{SP4}) = \left(\frac{2r\eta\rho(\rho + \delta)}{\delta N}, b_{SP4}, \frac{b_{SP4}}{\rho + \delta}, \frac{\eta}{N}\right),$$

$$P_{SP5} = (S_{SP5}, b_{SP5}, q_{SP5}, k_{SP5}) = \left(\frac{2r\eta\rho(\rho + \delta)}{\delta N}, b_{SP5}, \frac{b_{SP5}}{\rho + \delta}, \frac{\eta}{N}\right),$$

where

$$q_{SP2} = \frac{\sqrt{N}\sigma - \sqrt{N\sigma^2 - 8r(\rho + \delta + \eta)\rho}}{2\sqrt{N}}$$

$$q_{SP3} = \frac{\sqrt{N}\sigma + \sqrt{N\sigma^2 - 8r(\rho + \delta + \eta)\rho}}{2\sqrt{N}}$$

Fig. 2 The five steady state points in the space (k, q, S)

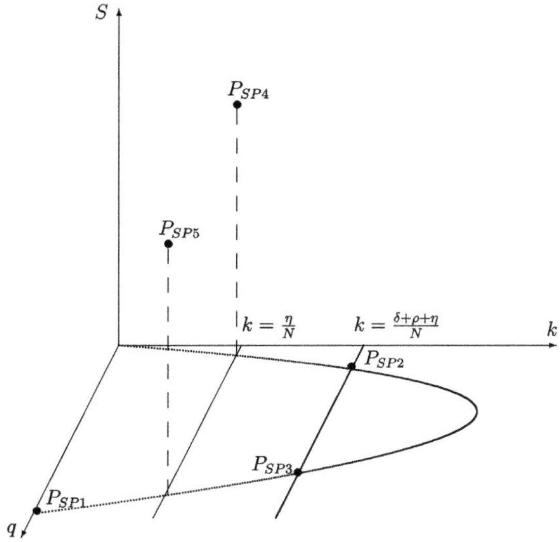

$$b_{SP4} = \frac{(\rho+\delta)(\sqrt{N}\sigma - \sqrt{N\sigma^2 - 8r\eta\rho})}{2\sqrt{N}}$$

$$b_{SP5} = \frac{(\rho+\delta)(\sqrt{N}\sigma + \sqrt{N\sigma^2 - 8r\eta\rho})}{2\sqrt{N}}$$

Proof Imposing stationarity on the R&D effort yields

$$k = \frac{q(\sigma - q)}{2r\rho} \tag{44}$$

which can be plugged into $\dot{q} = 0$ to obtain the following solutions:

$$q_{SP1} = \sigma; \qquad q_{SP2,3} = \frac{\sqrt{N}\sigma \mp \sqrt{N\sigma^2 - 8r(\rho+\delta+\eta)\rho}}{2\sqrt{N}} \tag{45}$$

with $q_{SP2,3} \in \mathbb{R}_+$ for $\sigma > \sqrt{8r(\rho+\delta+\eta)\rho/N}$. This in turns implies $k_{SP2,3} = (\delta + \rho + \eta)/N$. The corresponding state coordinates are $S = 0$, $b = 0$. On the other hand, if $b \neq 0$, $\dot{b} = 0$ in $k = \eta/N$ and plugging this expression into $\dot{k} = 0$, we obtain $q_{SP4,5}$. Consequently, $\dot{q} = 0$ yields $b_{SP4,5}$ and finally $\dot{S} = 0$ produces $S_{SP4,5}$. □

Figure 2 locates the five steady state points emerging under social planning in three dimensions, in the space (k, q, S). Note that the equilibrium points P_{SP4} and P_{SP5} entail a positive amount of pollution and therefore do not belong to the control plane. The existence of the fourth and fifth solutions depends on the fact that the dynamics of the output level (42) depends on b, denoting that the planner indeed takes into account the environmental impact of the production technology when choosing the output level.

The Jacobian matrix is:

$$J=\begin{pmatrix} -\delta & q & b & 0\\ 0 & \eta-Nk & 0 & -Nb\\ 0 & 1 & \rho+\delta+\eta-Nk & -N(q-\sigma)\\ 0 & 0 & \frac{1}{2r}(2q-\sigma) & \rho \end{pmatrix}. \tag{46}$$

By repeating a procedure analogous to the one carried out to produce Proposition 4, we can prove that:

Proposition 14 *P_{SP1}, P_{SP2}, P_{SP3}, P_{SP4} and P_{SP5} are saddle points.*

Next we are going to evaluate the profits and the social welfare levels at each equilibrium point.

$$\begin{aligned} \pi(P_{SP1}) &= 0,\\ \pi(P_{SP2}) = \pi(P_{SP3}) &= \frac{r}{N}\left(\rho^2-(\delta+\eta)^2\right),\\ \pi(P_{SP4}) = \pi(P_{SP5}) &= \frac{\eta r}{N}(2\rho-\eta). \end{aligned} \tag{47}$$

Proposition 15

1. *If $\rho\in(\delta+\eta,\infty)$, then the profits $\pi(P_{SP2})$, $\pi(P_{SP3})$, $\pi(P_{SP4})$ and $\pi(P_{SP5})$ are positive*;
2. *if $\rho\in(\delta+\eta,\delta+2\eta)$, then $\pi(P_{SP2})=\pi(P_{SP3})<\pi(P_{SP4})=\pi(P_{SP5})$.*

The social welfare associated to the steady states is computed as follows:

$$SW(P_{SPi}) = \pi(P_{SPi}) + \frac{q_{SPi}^2}{2} - S_{SPi}, \quad i=1,\dots,5, \tag{48}$$

and yields, respectively:

$$\begin{aligned} SW_1 &= SW(P_{SP1}) = \frac{\sigma^2}{2},\\ SW_2 &= SW(P_{SP2})\\ &= \frac{\sigma^2}{4} - \frac{\sigma\sqrt{\sigma^2N-8r(\rho+\delta+\eta)\rho}}{4\sqrt{N}} - \frac{r(\delta+\eta)(\delta+\eta+\rho)}{N},\\ SW_3 &= SW(P_{SP3})\\ &= \frac{\sigma^2}{4} + \frac{\sigma\sqrt{\sigma^2N-8r(\rho+\delta+\eta)\rho}}{4\sqrt{N}} - \frac{r(\delta+\eta)(\delta+\eta+\rho)}{N},\\ SW_4 &= SW(P_{SP4}) = \frac{\sigma^2}{4} - \frac{\sigma\sqrt{\sigma^2N-8r\eta\rho}}{4\sqrt{N}} - \frac{r\eta[2\rho^2+\delta(\eta+\rho)]}{\delta N},\\ SW_5 &= SW(P_{SP5}) = \frac{\sigma^2}{4} + \frac{\sigma\sqrt{\sigma^2N-8r\eta\rho}}{4\sqrt{N}} - \frac{r\eta[2\rho^2+\delta(\eta+\rho)]}{\delta N}. \end{aligned} \tag{49}$$

Proposition 16

1. *$SW_1 > SW_5 > SW_4$ and $SW_3 > SW_2$ over the whole admissible range of parameters;*
2. *if $\rho \in (\delta + \eta, \delta + 2\eta)$ and $\delta > 2\eta$, then $SW_5 > SW_3$.*

The steady state replicating the perfectly competitive outcome of the static model would look like the most desirable one, since the related level of social welfare exceeds all the remaining ones. However, it remains out of reach for all $b_0 > 0$.[12]

Additionally, there exists a subset of the admissible range of parameters in which the steady state P_{SP5} is both privately and socially preferable to all the steady state allocations arising from the open-loop Nash game among unregulated firms. With this in mind, we turn now our attention to the design of a Pigouvian tax/subsidy that may adjust firms' incentives so as to drive them to reproduce P_{SP1}.

5 Effects of a Pigouvian Taxation

In this section, a Pigouvian tax rate $\theta > 0$ is introduced, with taxation taking the form of a linear function of the environmental externality produced by the industry. We are going to investigate the role of θ as an incentive for firms to adjust their production activity in order to reach the social optimum, but we are not going to take into account the effect of the revenue from this tax. In fact, we may assume that the industry under consideration is negligible with respect to the whole economy, hence the income effect of the Pigouvian tax revenue does not alter the market demand structure.

Such a taxation affects each current-value Hamiltonian function, which now writes as:

$$\mathcal{H}_i(\cdot) = \left(\sigma - \sum_{j=1}^{n} q_j\right) q_i - r k_i^2 - \theta S + \lambda_i \dot{S} + \mu_{ii} \dot{b}_i + \sum_{j \neq i} \mu_{ij} \dot{b}_j . \tag{50}$$

As in Benchekroun and Long (1998, 2002), our objective here is to investigate whether this Pigouvian tax rate can be designed so as to reproduce the same social welfare level characterising the first best (that the planner himself would be, in general, unable to attain). This assumption leaves the FOCs (5) and (6) unchanged, whereas the adjoint equations (7) become:

$$\dot{\lambda}_i(t) = (\rho + \delta)\lambda_i(t) + \theta . \tag{51}$$

[12]That is, the equivalent of Remark 5 holds here. The proof of this fact follows the same lines as for the Cournot equilibrium of the open-loop game among firms. The details have been omitted for brevity.

The state-control system is as follows:

$$\dot{S} = Nbq - \delta S \tag{52}$$

$$\dot{b} = b(\eta - k) \tag{53}$$

$$\dot{q} = \frac{[(N+1)q - \sigma][\rho + \delta + \eta - k] + \theta b}{N+1} \tag{54}$$

$$\dot{k} = \rho k - \frac{q[\sigma - (N+1)q]}{2r} \tag{55}$$

As a consequence of taxation, firms' cost structure is modified to account for pollution, and therefore $\dot{q}$ depends on θb. As a consequence, also $\dot{k}$ depends on θb. Therefore,

$$\frac{dq}{dk} = \frac{2r[((N+1)q - \sigma)(\rho + \delta + \eta - k) + \theta b]}{(N+1)[2r\rho k - q(\sigma - (N+1)q)]} \tag{56}$$

and the slope of the Nash trajectory in the control plane becomes sensitive to pollution thanks to the Pigouvian tax rate.

Also in this case, multiple equilibrium points appear. Provided that the market is large enough, $\sigma > 2\sqrt{2(1+N)r\rho(\rho+\delta+\eta)}$, we obtain three steady states corresponding to no pollution:

$$\begin{aligned}
(S_1, b_1, q_1, k_1) &= \left(0, 0, \frac{\sigma}{N+1}, 0\right); \\
(S_2, b_2, q_2, k_2) &= \left(0, 0, \frac{\sigma - \sqrt{\sigma^2 - 8(1+N)r\rho(\rho+\delta+\eta)}}{2(1+N)}, \rho+\delta+\eta\right); \\
(S_3, b_3, q_3, k_3) &= \left(0, 0, \frac{\sigma + \sqrt{\sigma^2 - 8(1+N)r\rho(\rho+\delta+\eta)}}{2(1+N)}, \rho+\delta+\eta\right).
\end{aligned} \tag{57}$$

Moreover, as in the social planning case, two further equilibria with positive stocks of pollution exist:

$$\begin{aligned}
(S_4, b_4, q_4, k_4) = \Bigg(&\frac{2\eta Nr\rho(\delta+\rho)}{\delta\theta}, \frac{(\delta+\rho)(\sigma - \sqrt{\sigma^2 - 8\eta(1+N)r\rho})}{2\theta}, \\
&\frac{4\eta r\rho}{\sigma - \sqrt{\sigma^2 - 8\eta(1+N)r\rho}}, \eta\Bigg);
\end{aligned} \tag{58}$$

$$\begin{aligned}
(S_5, b_5, q_5, k_5) = \Bigg(&\frac{2\eta Nr\rho(\delta+\rho)}{\delta\theta}, \frac{(\delta+\rho)(\sigma + \sqrt{\sigma^2 - 8\eta(1+N)r\rho})}{2\theta}, \\
&\frac{4\eta r\rho}{\sigma + \sqrt{\sigma^2 - 8\eta(1+N)r\rho}}, \eta\Bigg).
\end{aligned} \tag{59}$$

The steady states (58) and (59) depend on the Pigouvian tax rate: in particular, notice that $q_4^\theta > q_5^\theta$ and that the associated steady state levels of pollution are decreasing in θ.

At this stage, it is worth carrying out a comparative analysis of the social welfare equilibrium levels again. Let q_i^θ, S_i^θ, π_i^θ and SW_i^θ be, respectively, the i-th steady state values in the present case, the levels of social welfare SW_i^θ is computed by the following formula:

$$SW_i^\theta = \frac{(Nq_i^\theta)^2}{2} + N\pi_i^\theta - S_i^\theta. \tag{60}$$

Evaluating (60) at the two steady states affected by the tax rate, we have:

$$\begin{aligned} SW_4^\theta &= \eta Nr\left[\eta\left(-\frac{8Nr\rho^2}{(\sigma - \sqrt{\sigma^2 - 8(1+N)\eta r\rho})^2} - 1\right)\right. \\ &\quad \left. + 2\rho\left(\frac{2\sigma}{\sigma - \sqrt{\sigma^2 - 8(1+N)\eta r\rho}} - \frac{\delta+\rho}{\delta\theta}\right)\right]; \\ SW_5^\theta &= \eta Nr\left[\eta\left(-\frac{8Nr\rho^2}{(\sigma + \sqrt{\sigma^2 - 8(1+N)\eta r\rho})^2} - 1\right)\right. \\ &\quad \left. + 2\rho\left(\frac{2\sigma}{\sigma + \sqrt{\sigma^2 - 8(1+N)\eta r\rho}} - \frac{\delta+\rho}{\delta\theta}\right)\right], \end{aligned} \tag{61}$$

so $SW_4^\theta > SW_5^\theta$ irrespective of all the parameter values.

Now we compare SW_4^θ with the maximum social welfare level that would be obtained under social planning case, i.e. $SW_1 = \frac{\sigma^2}{2}$, in order to derive the threshold values of the tax rate that allows to reach SW_1 under oligopolistic competition.

If we consider SW_4^θ as a function of $\theta \in (0, \infty)$, we can stress that it takes negative values when θ is close to zero:

$$\begin{aligned} &\lim_{\theta\to\infty} SW_4^\theta \\ &= \frac{2\eta Nr[4\eta r\rho(\eta + \eta N - \rho N) - (2\rho - \eta)\sigma(\sigma - \sqrt{\sigma^2 - 8(1+N)\eta r\rho})]}{(\sigma - \sqrt{\sigma^2 - 8(1+N)\eta r\rho})^2}. \end{aligned} \tag{62}$$

Moreover, SW_4^θ is strictly increasing, consequently admitting a horizontal asymptote, whose level is positive if $\rho \in (\frac{\eta}{2}, \frac{\eta(1+N)}{N})$. Call $K := K(\eta, r, \rho, N, \sigma)$ such a positive level.

If $K > \frac{\sigma^2}{2}$, then the optimal tax rate θ^* entailing the identity $SW_1 = SW_4^{\theta^*}$ is given by:

$$\theta^* = \frac{4\eta Nr\rho(\delta+\rho)}{\delta(2K - \sigma^2)}. \tag{63}$$

An analogous procedure can be carried out with SW_5^{θ}, where it can be easily ascertained that the tax rate θ^{**} such that $SW_1 = SW_5^{\theta^{**}}$ exceeds θ^* and the inequality with respect to the related externality levels is inverted, i.e. $S_4^{\theta^*} > S_5^{\theta^{**}}$. In other words, the tax rate that allows to reproduce the social welfare SW_1 is higher and corresponds to a higher level of output and pollution.

This seemingly counterintuitive fact relies on the identity leading to the value (63):

$$SW_{4,5}^{\theta} = \pi_{4,5}^{\theta} + CS_{4,5}^{\theta} - S_{4,5}^{\theta} = \frac{\sigma^2}{2}, \tag{64}$$

implying

$$\theta = \frac{4\eta N r \rho(\delta + \rho)}{\delta[2(\pi_{4,5}^{\theta} + CS_{4,5}^{\theta}) - \sigma^2]}. \tag{65}$$

Thus, inequality $q_4^{\theta} > q_5^{\theta}$ affects the denominator of the previous relation, because $\pi_4 + CS_4 > \pi_5 + CS_5$. Hence, the first best social welfare can be obtained by moving along two different paths: either with a larger quantity and a lower price but a higher externality level, or conversely with a smaller quantity and a higher price but a lower externality.[13]

5.1 An Interpretation of the Pigouvian Tax

The remarkable feature of the latter result is that, starting form a situation where the command optimum (point P_{SP1}) reproducing the perfectly competitive outcome is not, in general, attainable under planning except in the uninteresting case where the productive technology is completely green at the outset, it is nonetheless true that there exist an optimal stationary industrial policy whereby the regulator can drive profit-maximising firms to yield the same steady state welfare level associated with the first best, although of course at the price of a different surplus allocation and environmental externality. If the regulator is interested in the size of the total pie but not in the relative size of its slices, this is a price that he might well be willing to pay.

We can also provide further interpretation of θ: (51) implies that the presence of a Pigouvian taxation induces firms to shrink output levels as compared to the unregulated setting, whenever $\lambda_i(0) < 0$. That is, the policy maker, being aware of

[13]Note that the corresponding steady state profits are independent of θ:

$$\pi_{4,5}^{\theta} = \frac{\delta\sigma(\sigma \pm \sqrt{\Omega}) - 2\eta(N+1)r\Upsilon}{2\delta(N+1)^2}$$

where $\Omega = \sigma^2 - 8\eta\rho(N+1)r$ and $\Upsilon = 2\rho^2(N+1)N + \delta[\eta(N+1) + 2\rho N^2]$. There exist admissible parameter regions where the above profits are strictly positive.

the tradeoff between the price effect and the external effect implied by any change in output, is willing to accept an increase in price (as a result of the related higher degree of quasi-collusion) for the sake of reducing the environmental externality.

Furthermore, a direct comparison between (7) and (51) shows that the former admits the nil solution, whereas the latter does not, meaning that for every $\theta > 0$ no degenerate solution exists. In other words, no firm is allowed to attach a nil shadow price to the pollution stock, so all of them must take its accumulation dynamics into account. If we call $\lambda_i^*(t)$ the solution to (7) and $\lambda_i^{\theta}(t)$ the solution to (51), and assume equal initial conditions, i.e. in both cases the i-th firm assigns the same shadow price to pollution at time $t = 0$, solving the two equations leads us to deduce that θ corresponds to the following ratio:

$$\theta = (\rho + \delta)\frac{\lambda_i^{\theta}(t) - \lambda_i^*(t)}{e^{(\rho+\delta)t} - 1}. \tag{66}$$

Equation (66) highlights that the Pigouvian tax not only compels all firms to take into account the pollution accumulation, but is also proportional to the difference between the shadow prices respectively with and without taxation at each instant of time. Substantially, it provides an instantaneous measure of how much more weight firms are obliged to attribute to pollution when they are taxed.

6 Concluding Remarks

We have revisited the issue of the incentive for firms to carry out R&D efforts aimed at introducing environmental-friendly technologies. Contrary to the acquired view establishing that such an incentive is lacking due to the fact that firms fail to internalise the environmental externality, the dynamic approach we have adopted in the foregoing analysis shows that firms do have an R&D incentive in this direction throughout the game, although it may indeed vanish in one specific steady state, which portrays the equilibrium outcome of the corresponding static game. Such an incentive has no altruistic nature, being associated with a quasi collusive decision on output levels whereby any environmentally-oriented R&D is accompanied by a price increase.

Moreover, we have investigated the behaviour of the model under the assumption that a benevolent planner controls production and R&D, showing that the perfectly competitive outcome with marginal cost pricing and a totally clean technology is one of the possible steady states of the system, but is feasible only if initial conditions are such that the environmental externality is not an issue from the very outset.

Yet, as a (partial) remedy, we have found that there exists a feasible stationary Pigouvian tax rate able to induce profit-maximising firms to follow a path leading to the very same aggregate steady state welfare as in the first best.

The foregoing analysis can be extended in several directions, to examine feedback solutions, the implications of international trade with transboundary pollution and uncertainty affecting both the accumulation of pollution and the R&D outcome, all of these issues to be nested into a general equilibrium approach. These extensions are left for future research.

Acknowledgements We thank Hassan Benchekroun, Guido Cozzi, Aart de Zeeuw, Christoph Grimpe, Morton Kamien, Paul Klemperer, Arkady Kryazhimskiy, George Leitmann, Emmanuel Petrakis, Vladimir Veliov, Franz Wirl and the audience at IIOC'09 (Boston), the 14th Symposium of ISDG (Banff, 2010) and the workshop on Economic Growth and the Environment (TU Wien, Dec. 8, 2011) for useful comments and discussion. The usual disclaimer applies.

Appendix: List of Symbols

- N: number of oligopolistic firms;
- $q_i(t)$: i-th firm's output at time t;
- $Q_{-i}(t)$: aggregate output of all firms except the i-th one at time t;
- $Q(t)$: aggregate output of all firms at time t;
- $k_i(t)$: R&D effort level of the i-th firm at time t;
- $p(t) = a - Q(t)$: inverse market demand function at time t, where $a > 0$ is the related reservation price;
- $C_i = cq_i(t)$: production cost function for the i-th firm at time t, where $c > 0$ is the related marginal cost;
- $b_i(t)$: marginal contribution of the i-th firm to the stock of pollution at time t;
- $S(t)$: aggregate stock of pollution at time t;
- $\Gamma_i(t) = rk_i(t)^2$: R&D cost function for the i-th firm at time t, weighted by the constant $r > 0$;
- $\pi_i(t) = (p(t) - c)q_i(t) - \Gamma_i(t)$: profit function for the i-th firm at time t;
- $\eta > 0$: regeneration rate for the marginal contribution of firms over time;
- $\delta > 0$: decay rate for the stock of pollution over time;
- $\rho > 0$: intertemporal discount rate of the market;
- $\sigma = a - c > 0$: market dimension parameter;
- $\lambda_i(t)$, μ_{ii}, μ_{ij}: shadow prices attached by firms to all the dynamic constraints of the model;
- CS: consumer surplus;
- SW: social welfare.

References

Amacher, G. S., Koskela, E., & Ollikainen, M. (2004). Environmental quality competition and eco-labelling. *Journal of Environmental Economics and Management*, *47*, 284–306.

André, F. J., Gonzàlez, P., & Porteiro, N. (2009). Strategic quality competition and the Porter hypothesis. *Journal of Environmental Economics and Management*, *57*, 182–194.

Antweiler, W., Copeland, B. R., & Taylor, M. S. (2001). Is free trade good for the environment? *American Economic Review*, *91*, 877–908.

Bartz, S., & Kelly, D. L. (2008). Economic growth and the environment: theory and facts. *Resource and Energy Economics*, *30*, 115–149.

Benchekroun, H., & Chaudury, A. R. (2011). Environmental policy and stable collusion: the case of a dynamic polluting oligopoly. *Journal of Economic Dynamics and Control*, *35*, 479–490.
Benchekroun, H., & Long, N. V. (1998). Efficiency inducing taxation for polluting oligopolists. *Journal of Public Economics*, *70*, 325–342.
Benchekroun, H., & Long, N. V. (2002). On the multiplicity of efficiency-inducing tax rules. *Economics Letters*, *76*, 331–336.
Bottega, L., & De Freitas, J. (2009). Public, private and nonprofit regulation for environmental quality. *Journal of Economics and Management Strategy*, *18*, 105–123.
Bovenberg, A. L., & de Mooij, R. A. (1997). Environmental tax reform and endogenous growth. *Journal of Public Economics*, *63*, 207–237.
Chiou, J.-R., & Hu, J.-L. (2001). Environmental research joint ventures under emission taxes. *Environmental and Resource Economics*, *21*, 129–146.
Copeland, B. R., & Taylor, M. S. (1994). North-South trade and the environment. *Quarterly Journal of Economics*, *109*, 755–787.
Copeland, B. R., & Taylor, M. S. (2004). Trade, growth, and the environment. *Journal of Economic Literature*, *42*, 7–71.
Damania, D. (1996). Pollution taxes and pollution abatement in an oligopoly supergame. *Journal of Environmental Economics and Management*, *30*, 323–336.
Dasgupta, P. (2007). Commentary: the Stern Review's economics of climate change. *National Institute Economic Review*, *199*, 4–70.
Downing, P. B., & White, L. J. (1986). Innovation in pollution control. *Journal of Environmental Economics and Management*, *8*, 225–271.
Dragone, D., Lambertini, L., & Palestini, A. (2010). Dynamic oligopoly with capital accumulation and environmental externality. In J. Grespo Cuaresma, T. Palokangas & A. Tarasyev (Eds.), *Dynamic systems, economic growth and the environment*, Berlin: Springer.
Greaker, M. (2006). Spillovers in the development of new pollution abatement technology: a new look at the Porter hypothesis. *Journal of Environmental Economics and Management*, *52*, 411–420.
Grossman, G. M., & Krueger, A. B. (1995). Economic growth and the environment. *Quarterly Journal of Economics*, *110*, 353–377.
Hart, R. (2004). Growth, environment and innovation—a model with production vintages and environmentally oriented research. *Journal of Environmental Economics and Management*, *48*, 1078–1098.
Itaya, J. (2008). Can environmental taxation stimulate growth? The role of indeterminacy in endogenous growth models with environmental externalities. *Journal of Economic Dynamics and Control*, *32*, 1156–1180.
Karp, L., & Livernois, J. (1994). Using automatic tax changes to control pollution emissions. *Journal of Environmental Economics and Management*, *27*, 38–48.
Klemperer, P. (2007). *Priorities in fighting climate change* (Editorial). VoxEU, 13 December (http://www.gqq10.dial.pipex.com/).
Lombardini-Riipinen, C. (2005). Optimal tax policy under environmental quality competition. *Environmental and Resource Economics*, *32*, 317–336.
Lutz, S., Lyon, T. P., & Maxwell, J. W. (2000). Quality leadership when regulatory standards are forthcoming. *Journal of Industrial Economics*, *48*, 331–348.
Milliman, S. R., & Prince, R. (1989). Firm incentives to promote technological change in pollution control. *Journal of Environmental Economics and Management*, *17*, 247–265.
Mohr, R. D. (2002). Technical change, external economies, and the Porter hypothesis. *Journal of Environmental Economics and Management*, *43*, 158–168.
Nordhaus, W. D. (2007). A review of the Stern Review on the economics of climate change. *Journal of Economic Literature*, *45*, 686–702.
Poyago-Theotoky, J. A. (2007). The organization of R&D and environmental policy. *Journal of Economic Behavior and Organization*, *62*, 63–75.
Scott, J. T. (1996). Environmental research joint ventures among manufacturers. *Review of Industrial Organization*, *11*, 655–679.

Stern, N. (2007). *The economics of climate change: the Stern Review*. Cambridge: Cambridge University Press.
Weitzman, M. L. (2007). A review of the Stern Review on the economics of climate change. *Journal of Economic Literature*, *45*, 703–724.

Part IV
Applications of Dynamic Systems to Energy Supply

Utmost Fear Hypothesis Explores Green Technology Driven Energy for Sustainable Growth

Chihiro Watanabe and Jae-Ho Shin

1 Introduction

Japan constructed a sophisticated co-evolutionary dynamism between innovation and institutional systems by transforming external crises into a springboard for new innovation. This can largely be attributed to the unique features of the nation to have a strong motivation to overcoming fear based on xenophobia and uncertainty avoidance as well as abundant curiosity, assimilation proficiency, and thoroughness in learning and absorption (Hofstedo 1991; Watanabe and Zhao 2006).

Such explicit dynamism was typically demonstrated by technology substitution for energy in the 1970s (Watanabe 1999). This accomplishment can largely be attributed to similar substitution efforts in the 1960s which resulted in technology

C. Watanabe (✉)
Department of Industrial Management, Tokyo Seitoku University, 1-7-13 Jujodai, Kita-ku, Tokyo 114-0033, Japan
e-mail: watanabe.c.pqr@gmail.com

C. Watanabe
Tokyo Institute of Technology, 2-12-1 Ookayama, Meguro-ku, Tokyo 152-8552, Japan

C. Watanabe
National University of Singapore, Engineering Drive 1, Singapore 117574, Singapore

C. Watanabe
University of Jyväskylä, 40014 Jyväskylä, Finland

C. Watanabe
International Institute for Applied Systems Analysis (IIASA), Schlossplatz 1, 2361 Laxenburg, Austria

J.-H. Shin
Korea Institute of S&T Evaluation and Planning (KISTEP), 68 Mabang-gil, Seocho-gu, Seoul 137-130, Korea
e-mail: jhshin@kistep.re.kr

J. Crespo Cuaresma et al. (eds.), *Green Growth and Sustainable Development*, Dynamic Modeling and Econometrics in Economics and Finance 14,
DOI 10.1007/978-3-642-34354-4_9,

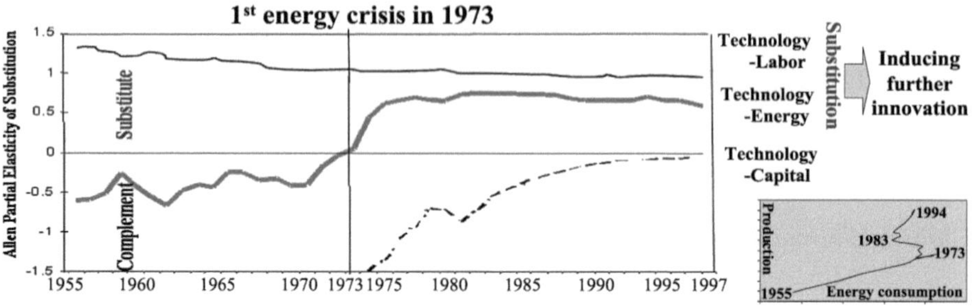

Fig. 1 Trends in technology substitution for production factors in the Japanese manufacturing industry *(1955–1997)—Allen partial elasticity of substitution*. Source: Watanabe (1999)

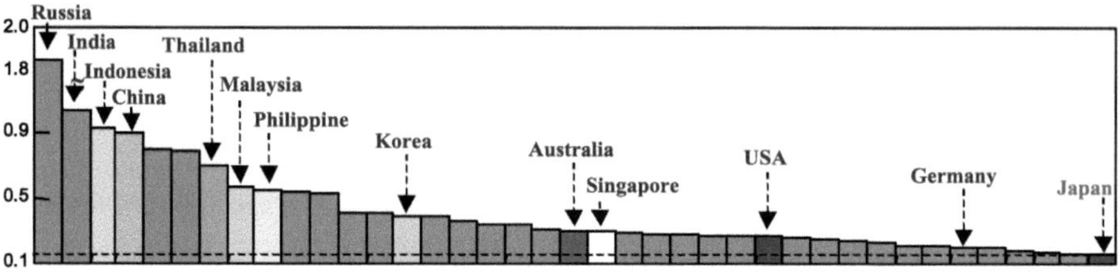

Fig. 2 Energy consumption per GDP in 40 countries *(2004)*. Source: IEA (2008a)

substitution for labor leading to world top-level labor saving and automation technologies (Shin 2009).

Supported by institutional systems for innovation, technology substitution for scarce resources functioned well in Japan typically in technology substituted for energy started from 1973 as demonstrated in Fig. 1 (Watanabe 1996; IEA 2008a). Figure 1 clearly illustrates complementary relationship between energy and technology before the first energy crisis in 1973 has changed to substitution after the crisis.

These cumulative technology substitution efforts subsequently enabled Japan to achieve a high-technology miracle in the 1980s.

Consequently, Japan demonstrates the world's highest energy efficiency as illustrated in Fig. 2. Furthermore, technology substitution for scarce resources led Japan demonstrates world top level of manufacturing technology.

Japan's unique institutional systems enabled conspicuous energy efficiency improvement in energy dependent industry in Japan as demonstrated in Fig. 3.

Thus, Japan constructed a sophisticated co-evolutionary dynamism between innovation and institutional systems by transforming external crises into a springboard for new innovation as illustrated in Fig. 4 which can largely be attributed to Japan's unique features of the nation to have (i) a strong motivation to overcoming fear based on xenophobia and uncertainty avoidance, (ii) while abundant curiosity, assimilation proficiency, and thoroughness in learning and absorption (Hofstedo 1991; Watanabe and Zhao 2006).

This unique institutional system led to high level of MPT (Marginal Productivity of Technology) leveraging conspicuously high level of elasticity of technology sub-

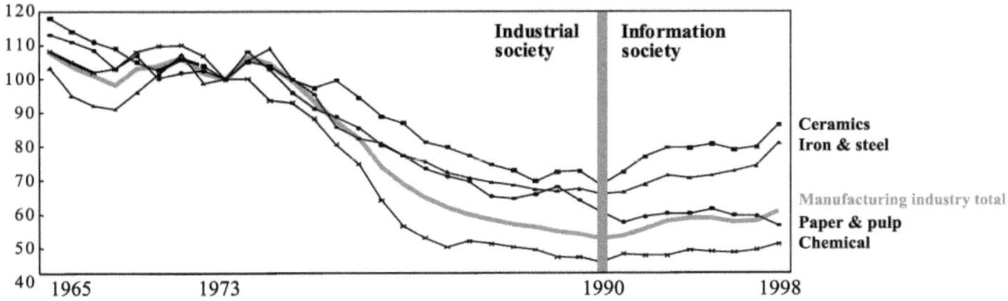

Fig. 3 Trend in unit energy consumption in the Japanese manufacturing industry *(1965–1998)—Index: 1973 = 100*

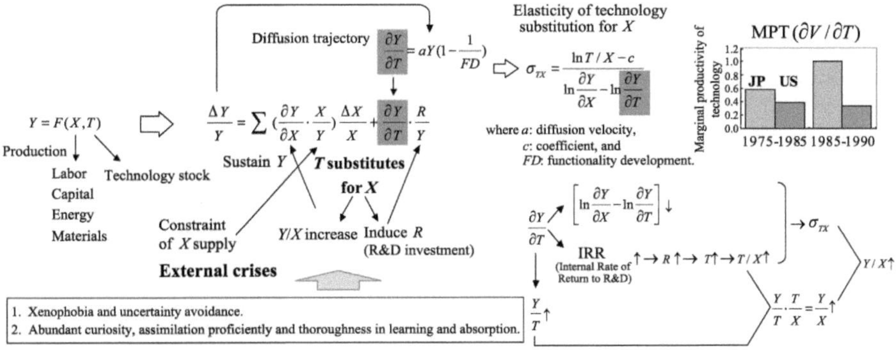

Fig. 4 Japan's notable dynamism in transforming external crises to a springboard for new innovation

stitution for constrained production factor X (T/X) leading to a shift from energy to technology (T/E) in the face of the energy crises, and productivity of technology increase (Y/T) which generate notable energy productivity as a multiplier effect of these accomplishment $(Y/E = (T/E) \times (Y/T))$. Enhanced energy productivity relaxed the energy constraints and enabled sustainable growth which again induced higher MPT leading to constructing a virtuous cycle between foregoing improvement as illustrated in Fig. 4.

While the dramatic increase in oil prices in the mid-2008 has signaled the possibility of a paradigm shift to a post-oil society, and not a few works demonstrated entrepreneurial strategies toward such a society (e.g., Bell 1973), none has identified the possible impacts of the sequel that such utmost fear ever experienced might provide even after overcoming the fear.

By applying Japan's notable dynamism, this paper attempted to identify a possible inducement that utmost fear may transform crises into a springboard for new innovation.

In line with the increasing significance of production, diffusion and consumption integration, and subsequent significance of the gratification of consumption for constructing a sustainable development dynamism, by applying a habit persistence hypothesis (Modigliani 1968) in which utmost gratification of consumption plays a decisive role in consumption behavior, an utmost fear hypothesis was developed. Given the unique features of the Japanese nation to have a strong motivation to overcoming xenophobia and uncertainty avoidance, Japan's innovation endeavor is very sensitive to such an utmost fear.

An empirical analysis based on the experience of a dramatic increase in oil prices was then conducted assuming photovoltaic solar cell (PV) development which is anticipated to play a leading role as technology-driven energy substituting for fossil energy (Watanabe et al. 2002) since it is a sophisticated renewable energy generation system by converting solar radiation into direct current electricity using semiconductors and transforming customers into producers of eco-friendly energy.

Despite a locational disadvantage as a mid-latitude country, Japan has taken a leading role in world PV development. Exceeding the level of the US in 1999, Japan maintained the world leading position in PV development before transferring this position to Germany in 2007 and then to China. Three main factors have been critical to the successful development of PV technology in Japan (Watanabe et al. 2002). First, like semiconductors, PV technology is central to a complex web of related technologies and can therefore benefit greatly from learning effects. Second, because of the interdisciplinary nature of PV development, technology spillover benefits are high, in turn further stimulating learning effects. Third, because of standalone, flexibility with respect to size, application and portability as well as multi-generational technological development, PV technology incorporates self-propagating development in its functionality similar to mobile phone (Watanabe et al. 2009). All corresponds to Japan's explicit learning and assimilation ability based on abundant curiosity, assimilation proficiency, and thoroughness in learning and absorption. Consequently, PV can be considered a crystal of institutional innovation suits to Japan's institutional systems (Watanabe et al. 2011).

On the basis of an empirical analysis on the development trajectory in Japan's PV development over the last 3 decades and also simulation analysis for the next two decades based on the experience of a dramatic increase in oil prices as US$147/b in July 2008, by utilizing the Bi-logistic growth model and Bass model, it was demonstrated that utmost fear plays a role similar to utmost gratification in consumption in leveraging a shift from resistance of innovation to supra-functionality development which incorporates social, cultural, aspirational, and emotional needs beyond economic value (McDonagh 2008) aiming at establishing a non-oil dependent resilient society.

In the current environment of simultaneous global economic stagnation, and given increasing concern regarding Japan's model for transforming a crisis into a springboard for new innovation, the foregoing analysis provides an important suggestion to firms with respect to their entrepreneurial strategy under open innovation in a post-oil society. Japan's March 11 catastrophe accelerates this demand.

Section 2 reviews utmost fear hypothesis leveraging a new innovation. Section 3 analyzes Japan's PV development trajectory. An empirical analysis aiming at demonstrating the utmost fear hypothesis is introduced in Sect. 4. Section 5 briefly summarizes the new findings, policy implications and the focus of the future works.

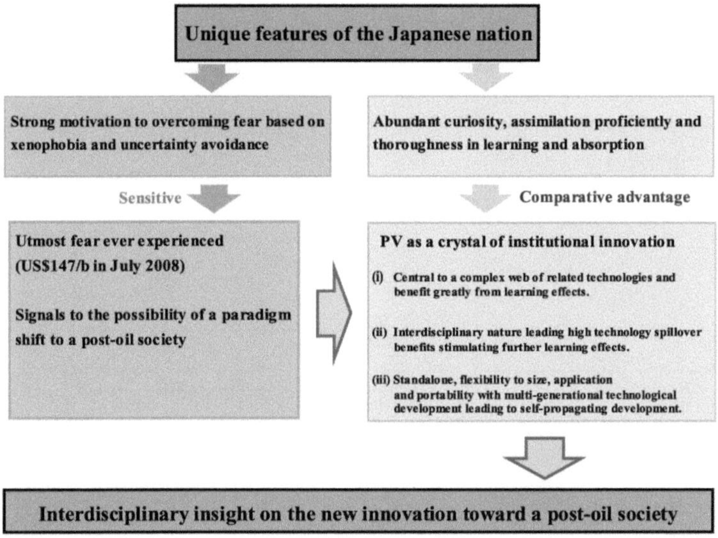

Fig. 5 Utmost fear hypothesis in Japan's PV development

2 Utmost Fear Hypothesis Leveraging a New Innovation

As reviewed in the preceding section, endogenous source of self-propagating functionality development (FD) can be expected by habit persistence hypothesis in that utmost gratitude of consumption plays a decisive role. Utmost fear like the dramatic increase in oil prices may play a similar role and leverages a shift from resistance of innovation (Bauer 1995) to supra-functionality (McDonagh 2008). Aiming at demonstrating this hypothetical view, an empirical analysis taking Japan's PV development in corresponds to the trend in oil prices was conducted.

Figure 5 illustrates the scheme of utmost fear hypothesis in Japan's PV development characterized by its unique feature of the nation sensitive to utmost fear ever experienced while incorporating explicit comparative advantage in PV development.

As mentioned earlier, Japan's institutional systems incorporate strong motivation to overcoming fear based on xenophobia and uncertainty avoidance which react sensitively to provide necessary countermeasures against signals to the possibility of a paradigm shift to a post-oil society. In order to provide necessary countermeasures against such signals, its unique features with abundant curiosity, assimilation proficiency and thoroughness in learning and absorption inevitably look for such institutional innovation as beneficiating from learning effects and spillover technology with self-propagating development nature. Since PV satisfies these requirements as a crystal of institutional innovation, high priority endeavor to its development should be natural consequence prompting an utmost fear hypothesis as illustrated in Fig. 5.

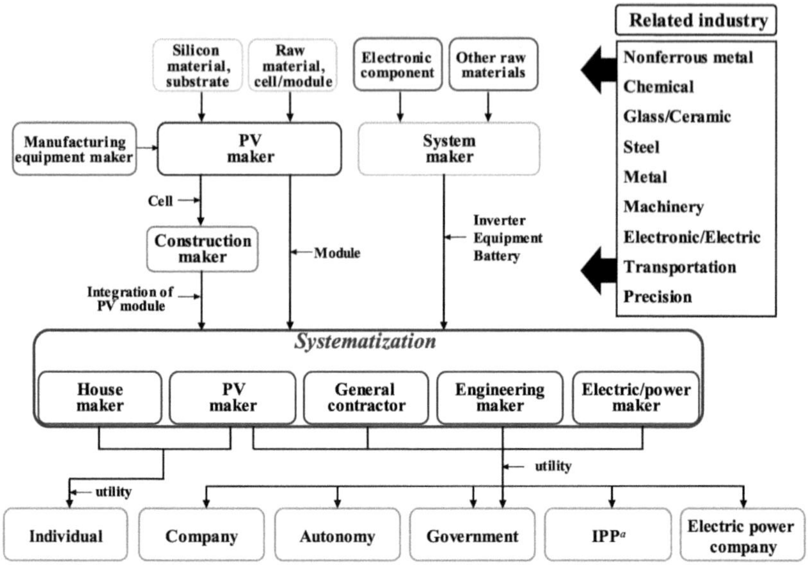

Fig. 6 Industrial network induced by PV development. [a]IPP: Independent Power Producer. Source: Authors' elaboration based on METI (Ministry of Economy and Industry) (2008)

3 PV Development Trajectory

Figure 6 illustrates the industrial networks induced by PV development in Japan. Figure 6 demonstrates that Japan's PV research institutes as well as makers of relevant industrial chain have established a comprehensive industrial network enabling Japan the leading PV development role in the global market.

Such a comprehensive industrial network can be attributed to government-industry joint work in overcoming Japan's fatal constraints in energy supply. The scarcity of natural conventional energy resources in Japan necessitates to accelerate the development and introduction of technology driven energy as PV.

Given the public nature of securing energy and risks and uncertainty indispensable for innovation, the Japanese government took initiative in endeavoring PV development and dissemination. Figure 7 demonstrates Japan's PV development trajectory over the last 3 decades.

Starting from the Sunshine Project undertaken immediately after the first energy crisis in 1973, successive efforts encompassing such broad approach as silicon (crystal and thin film), compound (mono crystal and poly crystal) and organic (pigment and nano-type) have been sustained aiming at developing next generation PV system. Joint work between the government, primarily national corporation NEDO (New Energy and Industrial Technology Development Organization) as well as government research institutes, and industry, primarily initiated by R&D consortium PVTECH, has been accelerated under the New Sunshine Program established in 1993 toward the development of the next generation PV system.

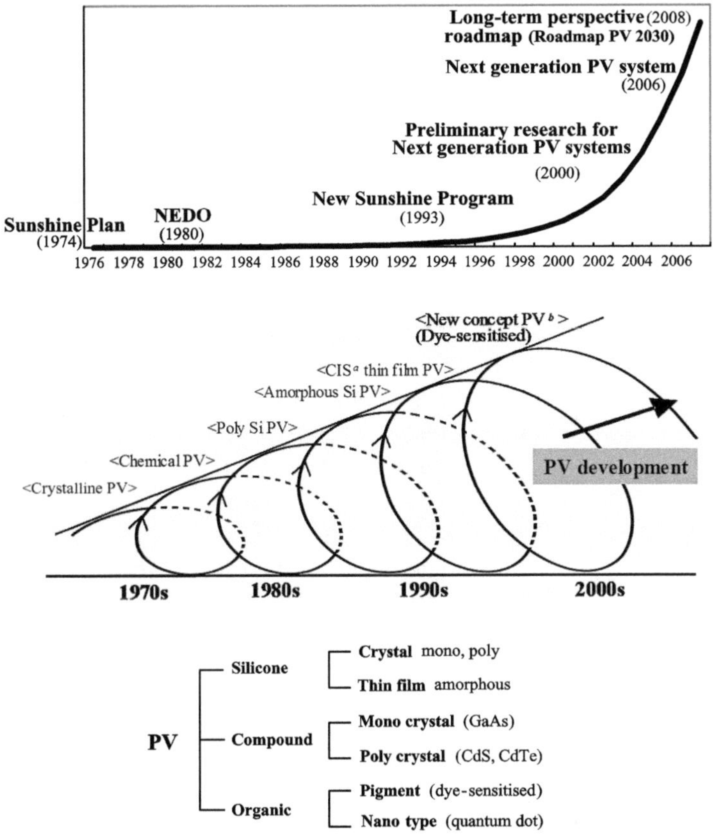

Fig. 7 Japan's PV development trajectory (1974–2008). [a]CIS: Copper-Indium-Diselenide compound PV. Source: Authors' elaboration based on METI (Ministry of Economy and Industry) (2009)

Table 1 and Fig. 8 review trends in PV development in the world over the period 1976–2007.

The Table and Figure demonstrate Japan's intensive endeavor in PV development as prospecting technology driven energy (Watanabe 1995, 1996; Watanabe et al. 2000; Watanabe and Asgari 2004). Japan exceeded its PV development level that of the US in 1999 and took the world leading position in this development in the world. However, Germany accelerated PV development from 2007 and since then the world leading position shifted from Japan to Europe and then to China.

On the basis of these trends in the world PV development, Table 2 and Fig. 9 demonstrate trends in cumulative PV development in the world over the period 1976–2007.

The Table and Figure demonstrate Japan's leading position in cumulative PV development since exceeding that of the US in 2000.

Table 1 Trends in PV development in the world *(1976–2007)*: MW

Year	Japan	USA	Europe	Others	Total
1976	0.01	0.32	0	0	0.33
1977	0.03	0.42	0	0	0.45
1978	0.06	0.84	0	0	0.90
1979	0.1	1.2	0	0	1.3
1980	0.3	2.5	0.3	0	3.1
1981	1.0	3.5	0.8	0	5.3
1982	2.1	5.2	1.4	0.1	8.8
1983	5.0	8.2	3.3	0.3	16.8
1984	8.9	8.0	3.6	0.8	21.3
1985	10.1	7.7	3.4	1.4	22.6
1986	12.8	7.1	4.0	2.3	26.2
1987	13.2	8.7	4.5	2.8	29.2
1988	12.8	11.1	6.7	3.0	33.6
1989	14.2	14.1	7.9	4.0	40.2
1990	16.8	14.8	10.2	4.7	46.5
1991	19.8	17.1	13.4	5.0	55.3
1992	18.8	18.1	16.4	4.6	57.9
1993	16.7	22.4	16.6	4.4	60.1
1994	16.5	25.6	21.7	5.6	69.4
1995	17.4	34.8	21.1	6.4	79.7
1996	21.2	38.9	18.8	9.8	88.7
1997	35.0	51.0	30.4	9.4	125.8
1998	49.0	53.7	33.5	18.7	154.9
1999	80.0	60.8	40.0	20.5	201.3
2000	128.6	75.0	60.7	23.4	287.7
2001	171.2	100.3	86.4	32.6	390.5
2002	251.1	120.6	135.1	55.1	561.9
2003	363.9	103.0	193.4	83.8	744.1
2004	602.0	139.0	314.0	140.0	1195.0
2005	832.6	154.0	470.0	302.0	1758.6
2006	927.5	201.6	678.3	714.0	2521.4
2007	920.0	266.1	1062.8	1484.0	3732.9

Source: PV Energy System Inc.: *PV News* (Monthly Issue)

Given the phased PV development with generations as demonstrated in Fig. 7, aiming at tracing its development by phases, on the basis of the trend in cumulative PV development in Japan, Table 3 analyzes the composition of Japan's PV development trajectory over the period 1976–2007 by utilizing the Bi-logistic growth model. Figure 10 illustrates the result of this analysis and demonstrates the PV development trajectory by decomposing phases 1 (Y_1) and 2 (Y_2) trajectories depending primarily

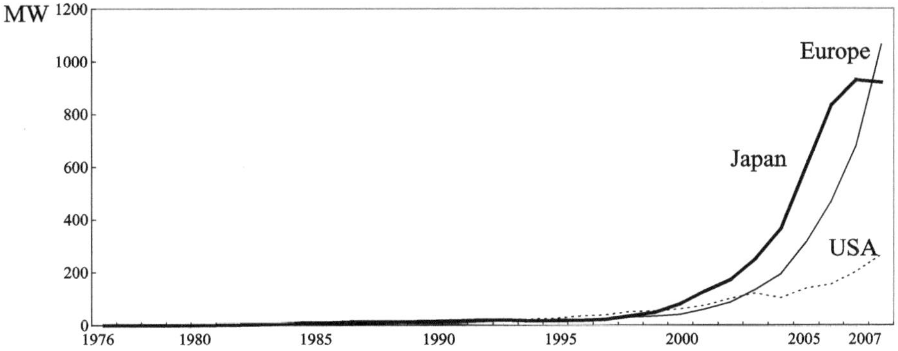

Fig. 8 Trends in PV development in Japan, the USA and Europe

on crystalline silicon based technology and on thin-film silicon technology, respectively.

Looking at the Table and Figure we note that Y_1 reaches its carrying capacity (upper ceiling) with 500 MW level. Y_1 changed from increase diffusion velocity to its decrease at inflection point in 2003 with 250 MW level. While Y_2 reaches its carrying capacity with 10,000 MW, Y_2 changes from increase diffusion velocity to its decrease at inflection point in 2009 with 5,000 MW level.

4 Utmost Fear Hypothesis

4.1 Optimal Functionality Development Dynamics

(1) Functionality Development Trajectory in Japan's PV Figure 11 demonstrates estimation of Japan's two phases of PV development trajectory over the period 1976–2007 analyzed by utilizing the Bi-logistic growth model as demonstrated in Table 3 and Fig. 10.

This phased development trajectory with generation suggests the self-developing mechanism of PV development as illustrated in Fig. 7.

(2) Optimal FD On the basis of the foregoing analyzes, optimal functionality development (FD) trajectory in Japan's PV development was analyzed. Figure 12 illustrates scheme of the identification of optimal FD trajectory (see details Appendix).

(3) Optimal FD Dynamics Leading to Supra-functionality As demonstrated in the preceding analysis, similar to mobile phones (Watanabe et al. 2009, 2011), PV incorporates self-propagating mechanisms in its development trajectory, and firms make every effort in maintaining sustainable FD in the "Innofumption (innovation through diffusion) dynamism," optimal FD trajectory should be endeavored in correspond to maximizing the gratification of consumption. With such understanding,

Table 2 Trends in cumulative PV development in the World *(1976–2007)*: MW

Year	Japan	USA	Europe	Others	Total
1976	0.01	0.32	0	0	0.33
1977	0.04	0.74	0	0	0.78
1978	0.10	1.58	0	0	1.68
1979	0.2	2.8	0	0	3.0
1980	0.5	5.3	0.3	0	6.1
1981	1.5	8.8	1.1	0	11.4
1982	3.6	14.0	2.5	0.1	20.2
1983	8.6	22.2	5.8	0.4	37.0
1984	17.5	30.2	9.4	1.2	58.3
1985	27.6	37.9	12.8	2.6	80.9
1986	40.4	45.0	16.8	4.9	107.1
1987	53.6	53.7	21.3	7.7	136.3
1988	66.4	64.8	28.0	10.7	169.9
1989	80.6	78.9	35.9	14.7	210.1
1990	97.4	93.7	46.1	19.4	256.6
1991	117.2	110.8	59.5	24.4	311.9
1992	136.0	128.9	75.9	29.0	369.8
1993	152.7	151.3	92.5	33.4	429.9
1994	169.2	177.0	114.2	39.0	499.4
1995	186.6	211.7	135.3	45.4	579.0
1996	207.8	250.6	154.1	55.2	667.7
1997	242.8	301.6	184.5	64.6	793.5
1998	291.8	355.3	218.0	83.3	948.4
1999	371.8	416.1	258.0	103.8	1149.7
2000	500.4	491.0	318.7	127.2	1437.3
2001	671.6	591.4	405.0	159.8	1827.8
2002	922.7	712.0	540.1	214.9	2389.7
2003	1286.6	815.0	733.4	298.7	3133.7
2004	1888.6	954.0	1047.4	438.7	4328.7
2005	2721.2	1108.0	1517.4	740.7	6087.3
2006	3648.7	1309.6	2195.7	1454.7	8608.7
2007	4568.7	1575.7	3258.5	2938.7	12341.6

Original source: PV Energy System Inc.: *PV News* (Monthly Issue)

PV development trajectory under certain investment intensity (cost minimum) that maximizes utility function (utility maximum) leading to utmost gratification of consumption (FD maximum) was analyzed based on optimal theory.

Figure 13 compares trend in actual level of FD in Japan's PV development trajectory over the last decade with that of optimal level of FD estimated by the foregoing optimal dynamics analysis.

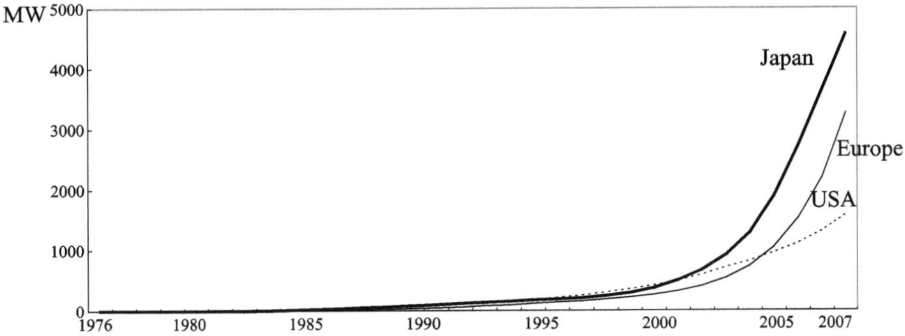

Fig. 9 Trends in cumulative PV development in Japan, the USA and Europe *(1976–2007)*: MW

Table 3 Estimation of Japan's PV diffusion by the Bi-logistic growth model *(1976–2007)*: MW. Model: $Y(t) = Y_1(t) + Y_2(t) = \frac{N_1}{1+b_1 \cdot e^{-a_1 t}} + \frac{N_2}{1+b_2 \cdot e^{-a_2 t}}$

Parameter	Estimate	t-value	adj. R^2
N_1	0.5×10^3	34.62	0.999
N_2	10.0×10^3	713.21	
a_1	4.58×10^{-1}	12.73	
b_1	26.0×10^5	3.60	
a_2	3.98×10^{-1}	41.34	
b_2	4.59×10^5	3.59	

Sub trajectory	Inflection point	Sub trajectory	Rate of obsolescence (monthly) ρ
Y_1	$t_1^{\#} = \frac{\ln b_1}{a_1}$	28.9 (2003)	$1/(28.9 \times 12) = 0.003$
Y_2	$t_2^{\#} = \frac{\ln b_2}{a_2}$	34.5 (2009)	$1/(34.5 \times 12) = 0.002$

Japan's PV development trajectory over the last decade demonstrates this dynamics suggesting a possibility of follower (optimal level) substitutes for leader (actual level) in open innovation. Among new concept PV systems are expected to grow, the most promising is dye-sensitized solar cells. Their early-stage products will be introduced into markets in 2008.

In 2006, NEDO (New Energy and Industrial Development Organization) has started public offering of PV field test. Its main purpose is to encourage the improvement of the necessary performance and lower cost for the full-fledged diffusion of PV.

(4) Dynamism Leading to Supra-functionality Dynamism leading to supra-functionality of Japan's MP (mobile phone), Web and PV can be demonstrated in Fig. 14. This Figure suggests that new FD frontier (e-mail transmission, Really

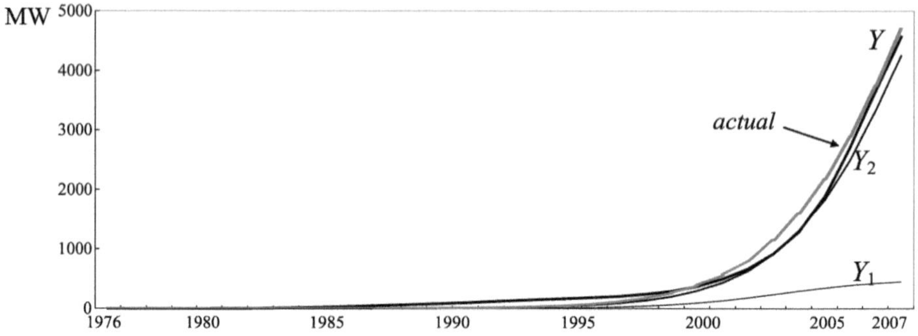

Fig. 10 Trends in diffusion trajectory of PV development in Japan *(1976–2007)*: MW

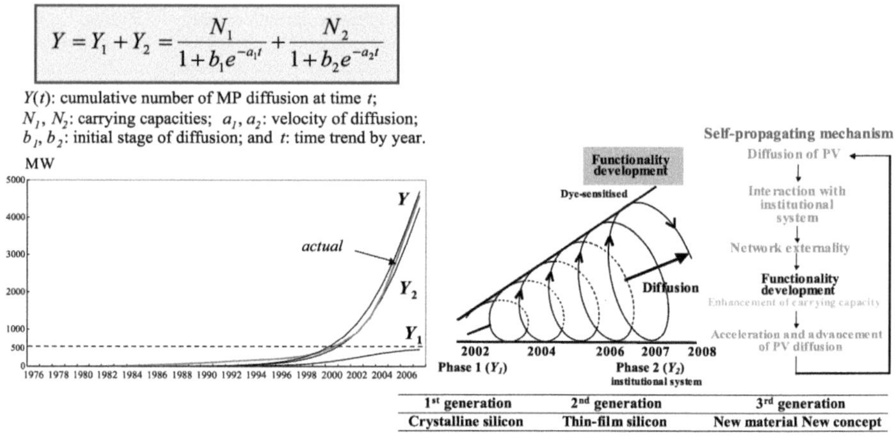

Fig. 11 Estimation of Japan's PV development trajectory by the Bi-logistic growth model *(1976–2007)*: MW

Simple Syndication (RSS), and next generation PV system (NGPVS), respectively) which instills in users an "exciting story on their own initiatives as heroes/heroines" thrills them with gratification beyond economic value. Sky Walker has incorporated new social, cultural and aspirational value of MP. Similar to MP, Web and PV also demonstrates that RSS (Really Simple Syndication) and NGPVS (Next Generation PV System) incorporate new FD frontier depicting "exciting story on their own initiatives as heroes/heroines."

Based on this Figure, MP e-mail transmission by Sky Walker suggesting supra-functionality substituted for resistance to new innovation and exploring new FD frontier leads to instill customers through mew communication community. Web also demonstrates supra-functionality through RSS 2.0, which enabling new FD frontier to encourage user participation embraced by Web publisher.

Similarly, the concept of supra-functionality in PV can be illustrated in Fig. 15. This Figure demonstrates that in line with the advancement of technology, Japan's

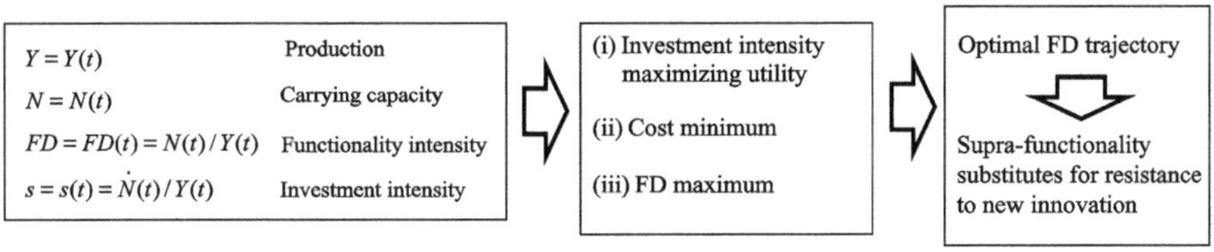

Fig. 12 Scheme of the identification of optimal FD trajectory

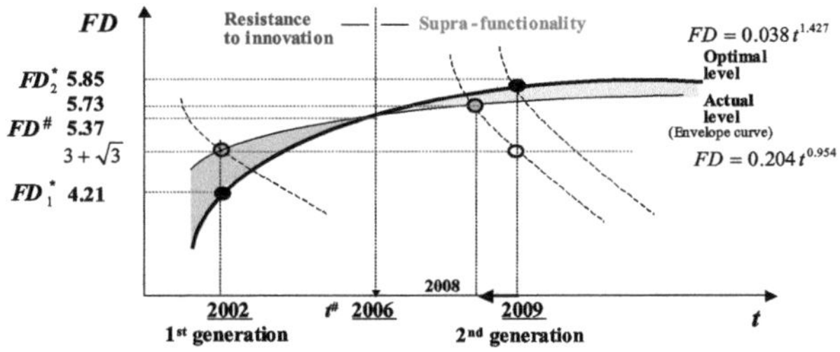

Fig. 13 Comparison of optimal and actual levels of FD in Japan's PV development trajectory *(1976–2007)*. [a]FD[#]: Utmost FD level: Level of FD at its emergence (Rogers, Mahajan and Moore). [b]Trajectory that satisfy investment intensity maximizing utility, cost minimum and FD maximum and depicted by the following equations (see Appendix) (Watanabe et al. 2011): $FD^* = \frac{a}{2\cdot(a+\rho)\cdot\rho}\cdot(-a+(a^2+4\cdot\frac{(a+\rho)\cdot\rho\cdot(a+1)}{a})^{1/2})$ where a and ρ: velocity of diffusion and discount coefficient (rate of obsolescence of technology), respectively. $Y_1(a,\rho) = Y_1(0.46, 0.003)$ and $Y_2(a,\rho) = Y_2(0.40, 0.002)$ (see Table 3)

PV development trajectory has been shifting from suppliers initiative to users initiative, and from economic value to social, cultural and Aspirational value correspondingly. Consequently, supra-functionality substituted for resistance to innovation in 2006 and stimulated by preceding innovation, new FD frontier was incorporated in PV in 2006 instilling users "exciting story," similar to Sky Walker in MP.

4.2 PV Development Inducement Against Utmost Fear

(1) Trends in Oil Prices ***(1972–2008)*** International oil prices demonstrated its peak level US$40/b in 1980 as a consequence of the 2nd oil crisis in 1979 as demonstrated in Fig. 16. While they changed to declining trend due to glut circumstances in the 1980s and the 1990s, they changed to dramatic increase from 2004. They recorded historical highest level as US$137/b in mid-2008 and changed to decline as demonstrated in Fig. 17.

(2) Prospects of Oil Prices ***(2009–2030)*** While international oil prices changed to decline with the peak in mid-2008, they are anticipated to change to increasing trend again in long run. IEA estimated that they will increase to 200 US$/b in 2030 as illustrated in Fig. 18 and Table 4.

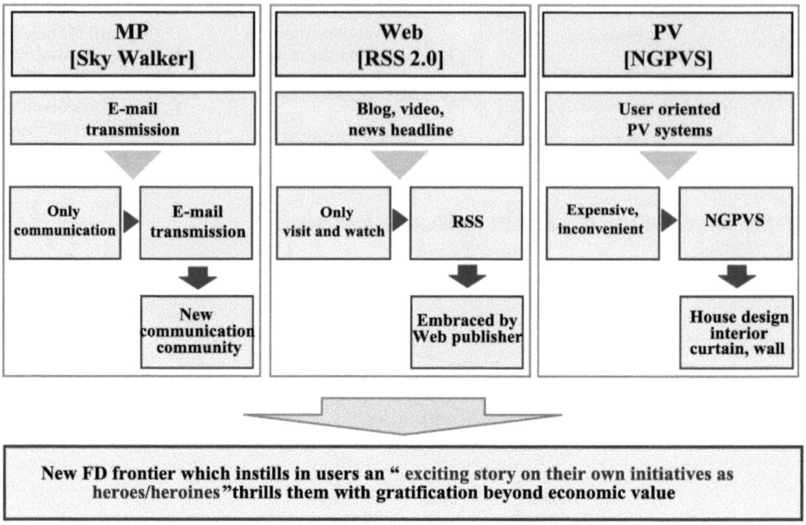

Fig. 14 Dynamism leading to supra-functionality of MP, Web and PV in Japan

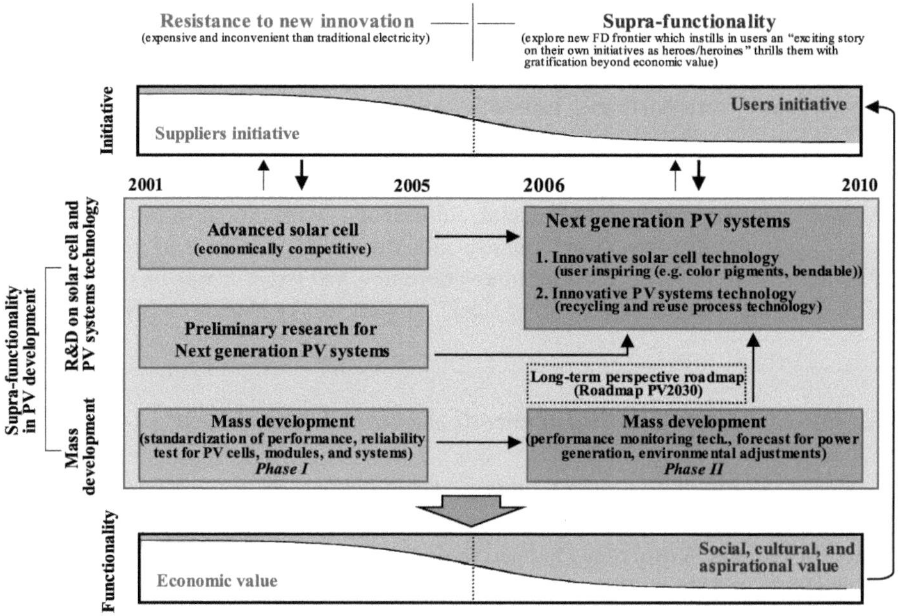

Fig. 15 Supra-functionality in PV development

(3) Possible PV Development Trajectories ***(1976–2030)*** Given the sensitive nature of PV development as an oil-alternative energy sensitive to oil prices, such an increase in oil prices inevitably accelerates PV development as has been broadly

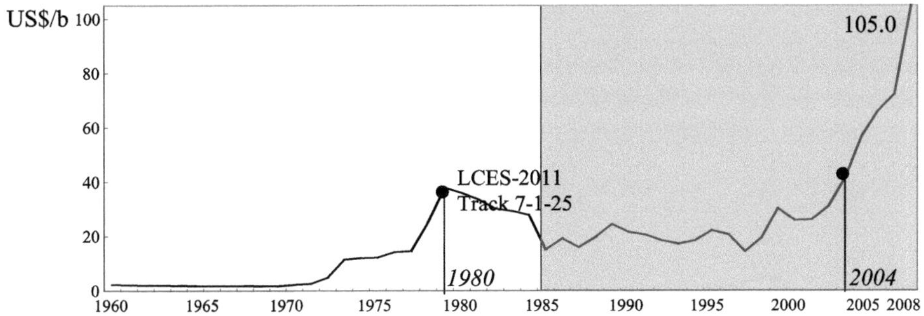

Fig. 16 Trend in oil prices *(1960–2008)*: current prices (US$/b). [a]Prices in 2008 is based on the average of monthly statistics between January and November 2008. Source: CIF import prices over the period 1960–1975 and International oil prices by WTI (West Texas Intermediate) (IEA 2008b) over the period 1976–2008

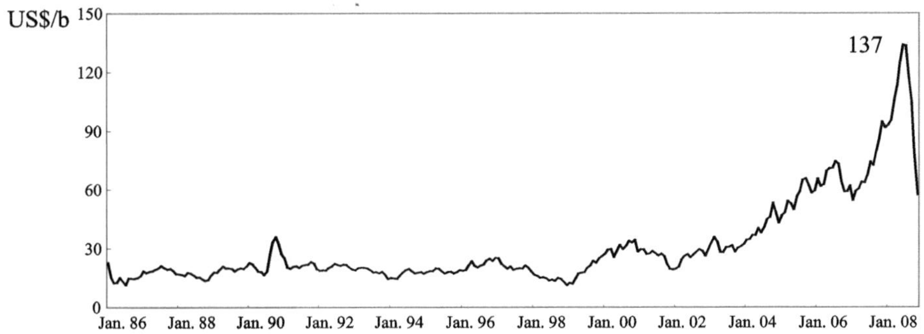

Fig. 17 Trend in oil prices *(Jan. 1986–Nov. 2008)*: current prices (US$/b). Source: International oil prices by WTI (West Texas Intermediate)

demonstrated by dramatic increase in PV development after the dramatic hike in oil prices in mid-2008.

Based on this correlation, Fig. 19 provides possible scenario of Japan's PV development toward 2030. On the basis of the estimate of PV development trajectory over the period 1976–2007 by means of Bi-logistic growth model as demonstrated in Table 3, PV development estimate scenario over the period 2008–2030 induced by oil prices increase were estimated with 20%, 30%, 50%, and 60% higher increase than the estimate by the Bi-logistic growth model without taking into account of the utmost fear effect.

On the basis of the foregoing scenario, with the possible estimate of the increase in oil prices as 5 US$/b p.a. increase from 2009 as illustrated in Fig. 19 (this estimate is lower than that of IEA), comparative analysis of the inducing impacts of oil prices increase on the advancement of Japan's PV was attempted, Table 5 summarizes the result of the analysis.

Table 5 compares inducing impacts of oil prices increase on the advancement of PV between direct impact and comprehensive impacts with utmost fear. While

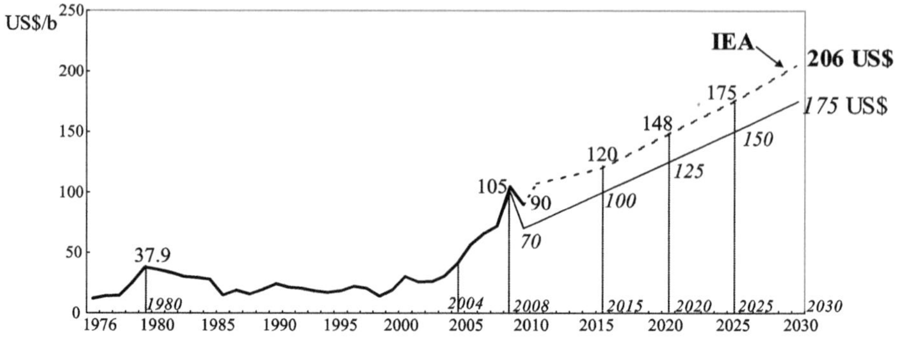

Fig. 18 Prospect of international oil prices *(1976–2030)*. Source: Author's estimation scenario based on World Energy Outlook 2008 (Watanabe 1996)

Table 4 Prospect of international crude oil prices *(2007–2030)*: US$/b

	2007	2010	2015	2020	2025	2030
Current prices	69.0	107.3	120.3	148.2	175.1	206.4
2007 fixed prices	69.0	100.0	100.0	110.0	116.0	122.0

Source: World Energy Forecast 2008 (IEA 2008a)

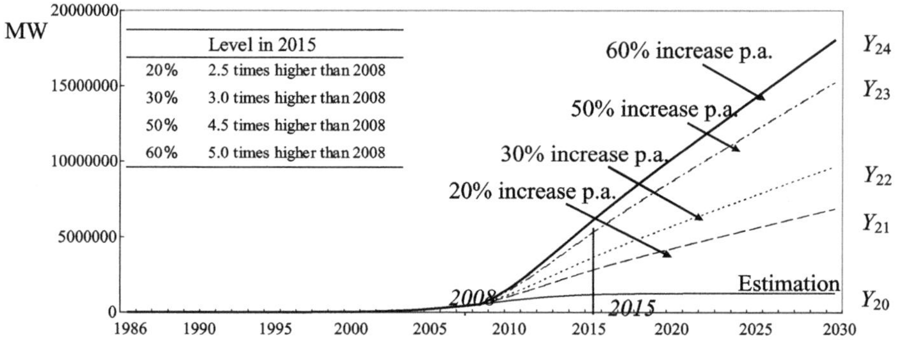

Fig. 19 Prospects of PV development in Japan *(1986–2030)*[a]. [a]Estimate based on the Bi-logistic growth model over the period 1976–2007; estimate scenario over the period 2008–2030 are 20%, 30%, 50%, and 60% p.a. higher increase than the estimate by the Bi-logistic growth model, respectively. Source: Authors' estimation scenario based on World Energy Outlook 2008 (IEA 2008a)

the former simply analyzes the correlation between oil prices and PV development, the latter analyzes the impacts of the "utmost fear ever experienced" by taking the impacts of the balance between the highest level and the level in respective year. Table 5 clearly indicates that comprehensive impacts with utmost fear demonstrate statistical significance.

Table 5 Comparison of the inducing impacts of oil prices increase on the advancement of Japan's PV *(1986–2015)*

Direct impact	adj. R^2	*DW*	*AIC*	*F*
$\ln Y_{20} = -6.380 + 4.395 D_1 \ln P + 4.450 D_2 \ln P + 4.251 D_3 \ln P + 2.298 D_4$	0.907	1.17	0.56	71.35
(−2.34) (4.90) (6.77) (6.88) (6.44)				
$\ln Y_{21} = -6.912 + 4.601 D_1 \ln P + 4.584 D_2 \ln P + 4.575 D_3 \ln P + 2.131 D_4$	0.904	0.97	5.92	69.42
(−2.32) (4.69) (6.38) (6.77) (5.46)				
$\ln Y_{22} = -7.179 + 4.704 D_1 \ln P + 4.652 D_2 \ln P + 4.737 D_3 \ln P + 2.048 D_4$	0.902	0.89	8.88	67.92
(−2.30) (4.57) (6.16) (6.67) (4.99)				
$\ln Y_{23} = -7.712 + 4.911 D_1 \ln P + 4.786 D_2 \ln P + 5.061 D_3 \ln P + 1.881 D_4$	0.898	0.75	15.05	64.49
(−2.23) (4.30) (5.72) (6.43) (4.14)				
$\ln Y_{24} = -7.979 + 5.014 D_1 \ln P + 4.853 D_2 \ln P + 5.222 D_3 \ln P + 1.797 D_4$	0.895	0.70	18.15	62.74
(−2.19) (4.17) (5.51) (6.31) (3.76)				
$\ln Y_{25} = -8.246 + 5.117 D_1 \ln P + 4.921 D_2 \ln P + 5.384 D_3 \ln P + 1.714 D_4$	0.892	0.66	21.23	61.04
(−2.15) (4.04) (5.34) (6.18) (3.40)				

Table 5 (*Continued*)

Comprehensive impacts with utmost fear	adj. R^2	*DW*	*AIC*	*F*
$\ln Y_{20} = 17.050 - 3.497 D_1 \ln(P_{max} - P) - 1.400 D_2 \ln P - 1.644 D_3 \ln(P_{max} - P) + 2.353 D_4$ (14.48)(−8.40) (−4.62) (−3.88) (7.17)	0.921	1.54	−4.32	85.065
$\ln Y_{21} = 14.881 - 2.814 D_1 \ln(P_{max} - P) - 0.929 D_2 \ln P - 1.044 D_3 \ln(P_{max} - P) + 2.669 D_4$ (13.01)(−7.42) (−3.40) (−2.78) (7.93)	0.943	1.35	−9.74	121.27
$\ln Y_{22} = 15.678 - 3.084 D_1 \ln(P_{max} - P) - 1.118 D_2 \ln P - 1.188 D_3 \ln(P_{max} - P) + 2.656 D_4$ (13.74)(−8.15) (−4.10) (−3.17)(7.91)	0.948	1.49	−9.89	132.40
$\ln Y_{23} = 17.273 - 3.624 D_1 \ln(P_{max} - P) - 1.495 D_2 \ln P - 1.476 D_3 \ln(P_{max} - P) + 2.628 D_4$ (14.60)(−9.24) (−5.28) (−3.79) (7.54)	0.952	1.69	−7.70	144.73
$\ln Y_{24} = 18.071 - 3.893 D_1 \ln(P_{max} - P) - 1.683 D_2 \ln P - 1.620 D_3 \ln(P_{max} - P) + 2.615 D_4$ (14.73)(−9.57) (−5.74) (−4.01) (7.24)	**0.952**	**1.75**	**−5.53**	**145.65**
$\ln Y_{25} = 18.868 - 4.163 D_1 \ln(P_{max} - P) - 1.872 D_2 \ln P - 1.763 D_3 \ln(P_{max} - P) + 2.601 D_4$ (14.70)(−9.79) (−6.10) (−4.18) (6.88)	0.952	1.78	−2.82	143.74

Y_{2i} ($i = 1 \sim 5$): cumulative stock of PV diffusion in phase 2 with extended estimation with annual increase rate of 20% (Y_{21}), 30% (Y_{22}), 50% (Y_{23}), 60% (Y_{24}) and 70% (Y_{25}), respectively; P: international oil prices (US$/bbl at current prices) by WTI (West Texas Intermediate) with extended estimation of 5US$/b p.a. increase from 2009; and D_i ($i = 1 \sim 3$): dummy variables with following classifications:

Dummy variables	1986–2003	2004–2008	2009–2015
D_1	1	0	0
D_2	0	1	0
D_3	0	0	1

D_4: 1994–2003 = 1, 2006–2012 = 1 and other years = 0 (comprehensive case)
D_4: 1994–2004 = 1, 2009–2012 = 1 and other years = 0 (direct impact case)

Table 5 then compare statistical significance between 6 scenario estimated with respect to possible accelerated PV development as a consequence of oil prices increase as estimated in Fig. 19 and indicates that 60% increase p.a. scenario with highest F value demonstrates statistically most significant.

These analyzes demonstrate that dramatic hike in oil prices induces accelerated PV development as a reaction to utmost fear and a dramatic increase in oil prices as US$137/b experienced in July 2008 induces dramatic acceleration of Japan's PV development as 5 times higher level than that of 2008 in 2015.

(4) Effects of Utmost Fear in Inducing PV Development On the basis of the result of the foregoing empirical analysis, effects of utmost fear in inducing PV development were analyzed.

PV production X can largely be attributed to the inducement of international oil prices P as follows:

$$X = F(P) \tag{1}$$

Given that g and ρ indicate initial increase rate of X and rate of obsolescence of cumulative stock of PV Y, respectively, Y can be depicted as follows:

$$Y \approx \frac{X}{g+\rho} \tag{2}$$

Provided that $g + \rho \equiv A$ is stable, (2) can be rewritten as follows:

$$Y \approx \frac{X}{A} \tag{3}$$

Taylor expansion of (1) to the first term

$$\ln X = a + b_1 \ln P \tag{4}$$

$$\ln AY = a + b_1 \ln P \tag{4'}$$

$$\ln Y = (a - \ln A) + b_1 \ln P \equiv a_1 + b_1 \ln P \tag{4''}$$

where a, b_1, and a_1 $(= a - \ln A)$: coefficients.

Following habit persistent hypothesis in consumption theory, comprehensive impacts of Y increase with utmost fear can be depicted as follows:

$$\ln Y = a_2 + b_2 \ln\left(P^* - P\right) \tag{5}$$

where P^*: utmost highest prices of oil; and a_1, b_1: coefficients.

Price elasticity to cumulative PV stock can be developed from (4′) and (5) as follows:[1]

[1]From (5), under $P^* \gg P$ condition,

$$\ln Y = a_2 + b_2 \ln P^*\left(1 - \frac{P}{P^*}\right) \approx a_2 + b_2 \ln P^* - b_2 \frac{P}{P^*} = a_2' - b_2' P$$

where $a_2' = a_2 + b_2 \ln P^*$, and $b_2' = \frac{b_2}{P^*}$

$$\frac{d \ln Y}{dP} = -b_2', \quad \frac{d \ln Y}{d \ln P} = \frac{d \ln Y}{dP} P = -b_2' P = -b_2 \frac{P}{P^*}$$

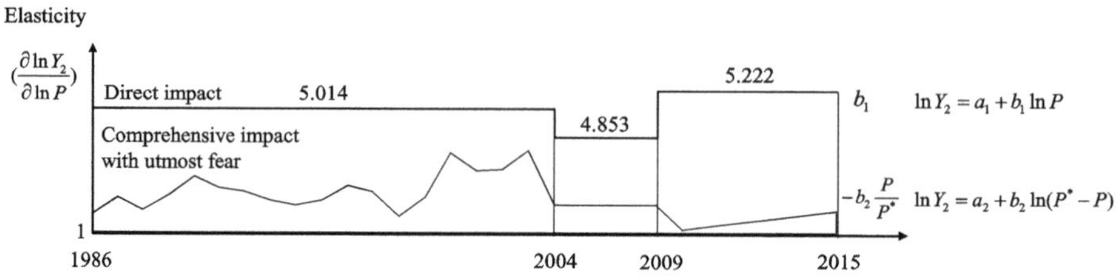

Fig. 20 Elasticity of oil prices to PV development in comprehensive impacts with utmost fear *(1986–2015)*

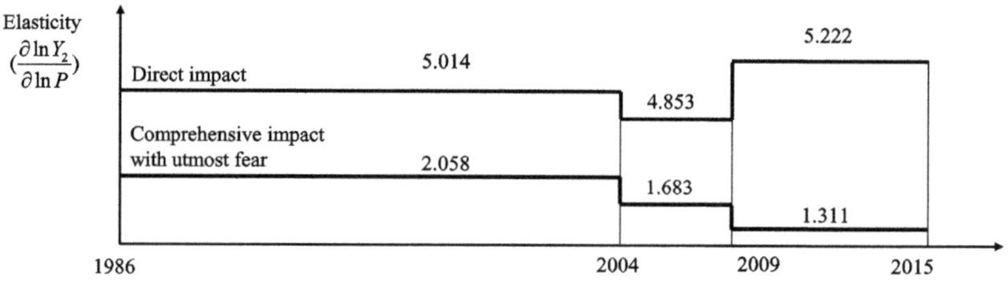

Fig. 21 Elasticity of oil prices to PV development in comprehensive impacts with utmost fear *(1986–2015)*

$$\varepsilon_1 \equiv \frac{\partial \ln Y}{\partial \ln P} = b_1 \quad \text{for direct impact} \tag{6}$$

$$\varepsilon_2 \equiv \frac{\partial \ln Y}{\partial \ln P} = -b_2 \frac{P}{P*} \quad \text{for comprehensive impacts with utmost fear} \tag{7}$$

On the basis of equations (6) and (7), both elasticity of oil prices to PV development in direct impact (DI) and comprehensive impacts with utmost fear (CIUF) were compared.

Figures 20 and 21 demonstrate the result of the comparison. Looking at the Figures we note that CIUF proves extremely lower elasticity of oil prices to PV development demonstrating consistent PV development independent from oil prices decrease (explicit ratchet effect).

(5) Impacts of Oil Prices Increase in Inducing PV Development Endeavors
Such inducement can be demonstrated by significant correlation between oil prices and number of PV endeavors measured by number of PV development projects. Such endeavors enable substitution of supra-functionality PV for resistance to its introduction.

Figure 22 demonstrates trends in number of PV endeavors by means of number of PV development projects appeared in monthly issue of PV News over the period 1997–2008.

Table 6 summarizes the result of the correlation analysis between oil prices and number of PV development in Japan and also in abroad over the period 1997–2008. Table 6 indicates statistically significance demonstrating that oil prices increase definitely induces PV development endeavors both in Japan and in abroad.

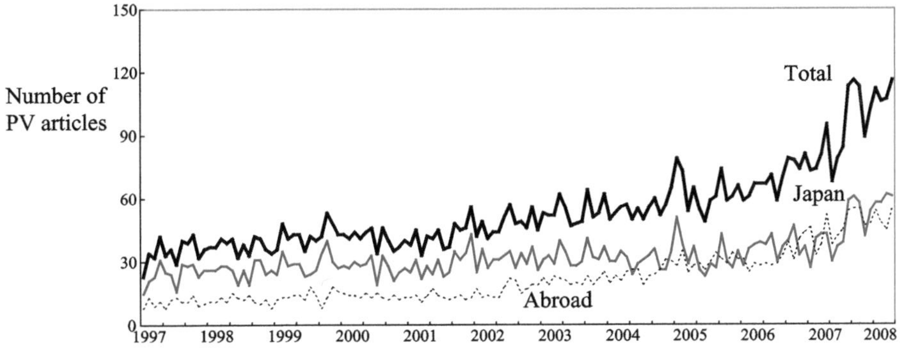

Fig. 22 Trends in number of PV endeavors *(Jul. 1997–Oct. 2008)*[a]. [a]Number of projects endeavoring to PV development introduced by PV News (see details in Appendix). Source: *PV News* (PV Energy System Inc., monthly issue)

Table 6 Impacts of oil prices increase in inducing PV development endeavors in Japan and abroad *(Jul. 1997–Oct. 2008)*: 3 months moving average

	adj. R^2	*DW*
$\ln N_{\text{Japan}} = 2.839 + 0.118 D_1 \ln P + 0.146 D_2 \ln P + 0.198 D_3 \ln P + 0.168 D_4$	0.881	1.47
(39.96) (5.14) (7.81) (12.37) (13.47)		
$\ln N_{\text{abroad}} = 1.010 + 0.467 D_1 \ln P + 0.533 D_2 \ln P + 0.600 D_3 \ln P + 0.183 D_4$	0.975	1.10
(13.36) (19.11) (26.90) (35.32) (13.77)		
$\ln N_{\text{total}} = 2.821 + 0.260 D_1 \ln P + 0.302 D_2 \ln P + 0.361 D_3 \ln P + 0.123 D_4$	0.977	1.39
(62.38) (7.79) (25.49) (35.51) (15.57)		

where N_{Japan}, N_{abroad}, and N_{total}: number of projects endeavoring to PV development in Japan, abroad, and World total, respectively introduced by *PV News*; P: international oil prices (US$/bbl at current prices) by WTI (West Texas Intermediate); and D_i (i = 1–4): dummy variables with following classifications:

Dummy variables	*Aug. 1997–Mar. 2002*	*Apr. 2002–Feb. 2007*	*Mar. 2007–Sep. 2009*
D_1	1	0	0
D_2	0	1	0
D_3	0	0	1

Prompted by these observations, the following analysis demonstrates PV development endeavors among Japan, abroad and total over the period 1997–2008:

$$N = Ae^{b_i D_i t} \quad (i = 1\text{–}3) \tag{8}$$

$$\ln N = \ln A + b_i D_i t \tag{8'}$$

where N: PV development endeavors; A and b_i: coefficients; D_i: dummy variable; and t: time trend.

Table 7 Impacts of PV development endeavors in Japan and abroad (Jul. 1997–Oct. 2008): 3 months moving average

	adj. R^2	DW
$\ln N_{Japan} = 3.117 + 0.0028D_1t + 0.0032D_2t + 0.0050D_3t + 0.176D_4$	0.883	1.40
(157.05) (5.10) (13.70) (25.06) (14.09)		
$\ln N_{abroad} = 2.202 + 0.008D_1t + 0.010D_2t + 0.012D_3t + 0.172D_4$	0.978	1.00
(112.94) (13.87) (43.89) (62.72) (14.01)		
$\ln N_{total} = 3.492 + 0.004D_1t + 0.006D_2t + 0.008D_3t + 0.121D_4$	0.973	1.06
(262.36) (10.44) (35.34) (56.87) (14.18)		

where N_{Japan}, N_{abroad}, and N_{total}: number of projects endeavoring to PV development in Japan, abroad, and World total, respectively introduced by *PV News*; and D_i ($i = 1$–4): dummy variables with following classifications:

Dummy variables	*Aug. 1997–Mar. 2002*	*Apr. 2002–Feb. 2007*	*Mar. 2007–Sep. 2009*
D_1	1	0	0
D_2	0	1	0
D_3	0	0	1

Table 8 Increase rate of PV development endeavors in Japan and abroad *(Jul. 1997–Oct. 2008)*: 3 months moving average

	D_1 (97/8–02/3)	D_2 (02/4–07/2)	D_3 (07/3–08/10)	D_3/D_2 increase
Japan	3.3	3.8	6.0	60%
Abroad	9.6	12	14.4	20%
Total	4.8	7.2	9.6	33%

Equation (8′) can be developed as following equation (9):

$$\frac{\Delta N}{N} = b_i D_i \tag{9}$$

Based on (9), increase rate of PV development endeavors can be identified. Table 7 demonstrates the significant impacts of PV development endeavors and dramatic increase trend.

On the basis of correlation analysis in Table 7, dramatic increase rate can be identified as summarized in Table 8.

Table 8 demonstrates a conspicuous increase rate of PV development endeavors and also this suggests that Japan's distinguished efforts for new innovation toward a post-oil society. Furthermore, Japan accomplished the highest increase rate of PV development endeavors compared to aboard and total.

Consequently, Japan's innovation toward PV development is further anticipated.

5 Conclusion

In light of the increasing concern regarding Japan's model for transforming a crisis into a spring board for new innovation in the current environment of simultaneous global economic stagnation and also a signal of the possibility of a paradigm shift to a post-oil society triggered by the dramatic increase in oil prices in mid-2008, this paper attempted to identify a new entrepreneurial strategy toward such a society by applying Japan's notable dynamism.

Given increasing concern on Japan's model for transforming a crisis into a springboard for new innovation particularly in the current environment of global economic stagnation, identification of innovation dynamism toward a post-oil society based on this approach is Japan's significant contribution to the global community.

Based on the review of Japan's notable dynamism in transforming a crisis into a springboard for new innovation and also the increasing significance of production, diffusion and consumption integration, utmost fear hypothesis leveraging the new innovation toward a post-oil society was examined by means of an empirical analysis on the development trajectory in Japan's PV development.

Noteworthy findings obtained include:

(i) Japan constructed a sophisticated co-evolutionary dynamism between innovation and institutional systems by transforming external crises into a springboard for new innovation.
(ii) This was typically demonstrated by technology substitution for energy in the 1970s enabling Japan to achieve a high-technology miracle in the 1980s.
(iii) This can be attributed to the unique features of the nation to have a strong motivation to overcoming fear based on xenophobia and uncertainty avoidance as well as abundant curiosity, assimilation proficiency, and thoroughness in learning and absorption.
(iv) Since the dramatic increase in oil prices has signaled the possibility of a paradigm shift to a post-oil society, a new entrepreneurial strategy toward such a society is strongly expected.
(v) By applying a habit persistence hypothesis in which utmost gratification of consumption plays a decisive role in consumption, an utmost fear hypothesis was demonstrated.
(vi) Utmost fear plays a similar role to utmost gratification in leveraging a shift from resistance of innovation to supra-functionality development aiming at establishing a non-oil dependent resilient society.

These findings suggest the following policy implications suggestive to firms with respect to their entrepreneurial strategy under open innovation in a post-oil society:

(i) Utmost fear plays a similar role to utmost gratification in leveraging a shift from resistance of innovation to supra-functionality development aiming at establishing a non-oil dependent resilient society.
(ii) Japan's notable model in transforming external crises into a springboard for new innovation should be broadly applied.

(iii) Technology substitution for constraints should be pursued in the scope of the integration between production, diffusion and consumption function.
(iv) Utmost functionality development should be endeavored aiming at supra-functionality substitution for resistance of innovation.
(v) Utmost fear hypothesis should be applied for leveraging new innovation toward a post-oil society.
(vi) PV development should be accelerated for new institutional innovation in a post-oil society.

Further works should focus on the establishment of introduction and application of utmost fear hypothesis in broader fields.

Appendix: Optimal Functionality Development in Response to Utmost Fear

(1) Model Construction MP (mobile phone), Web and PV (photovoltaic) in Japan,

$$Y(t) = C(t) + I(t) = (1 - s(t))Y(t) + s(t)Y(t) \Rightarrow \begin{array}{c} C(t) = (1 - s(t))Y(t) \\ \ln C(t) = \ln Y(t) + \ln(1 - s(t)) \end{array}$$

where $C(t)$: consumption, $I(t)$: investment, and $s(t)$: investment intensity ($I(t)/Y(t)$).

(i) Main Variables

				price	cost
$t \in [t_0, +\infty)$	Time on the infinite horizon				
$Y = Y(t)$	Production	*First phase variable*		ψ_2	$C_2 = \psi_2 \cdot Y$
$N = N(t)$	Carrying capacity				
$FD = FD(t) = \frac{N(t)}{Y(t)}$	Functionality development (FD)				
$\eta = \eta(t) = \frac{Y(t)}{N(t)} = \frac{1}{FD(t)}$	Production to carrying capacity	*Second phase variable*	$\Rightarrow \theta(t) \equiv FD(t) - 1$	ψ_1	$C_1 = \psi_1 \cdot \theta$
$s = s(t) = \frac{\dot{N}(t)}{Y(t)}$	Investment intensity (II)	*Control variable*			

(ii) System's Dynamics

$$\begin{cases} \dot{Y}(t) = a \cdot Y(t) \cdot (1 - \eta(t)) \\ \dot{\eta}(t) = a \cdot \eta(t) \cdot \left[1 - \eta(t) - \frac{s(t)}{a} \cdot \eta(t)\right] \end{cases}$$

Stationary level of FD

Stationary condition $\dot{\eta}(0) = 0$

$$\Rightarrow (1 - \eta_0 \cdot (1 + \frac{s_0}{a})) = 0$$

Here, $Y(t)$ represents GDP at time t.

Constant Levels of Investment Intensity (II)

Constraint $0 < s(t) = s(0) \equiv s_0 \leq A < 1$

Gratification of consumption sustaining stationary level of functionality development

It is necessary for accurate application of the Pontryaginmaximum principle.

If this constraint is satisfied, one can prove the existence result for the optimal control problem.

$$s_0 = a(\frac{1 - \eta_0}{\eta_0}) = a \cdot (\frac{1 - Y_0 / N_0}{Y_0 / N_0}) = a \cdot \frac{(N_0 - Y_0)}{Y_0} = a \cdot (FD_0 - 1)$$

(2) Optimal Control Problem for Functionality Development

$$\begin{cases} \theta(t) \equiv FD(t) - 1 \Leftrightarrow FD(t) = \theta(t) + 1 \Leftrightarrow \dot{FD}(t) = \dot{\theta}(t) \qquad \eta(t) = \frac{1}{FD(t)} = \frac{1}{\theta(t) + 1} \\ \dot{\theta}(t) = \dot{FD}(t) = s(t) - a \cdot (FD(t) - 1) = s(t) - a\theta(t) \\ \dot{Y}(t) = a \cdot Y(t) \cdot (\frac{\theta(t)}{\theta(t) + 1}) \end{cases}$$

(3) Utility Function (Integrated Logarithmic Consumption Index)[2)]

Consumption $C(t) = F(FD(t), Y(t), s(t)) = F(\theta(t), Y(t), s(t))$

$$U(\theta(t), Y(t), s(t)) = \int_0^{+\infty} e^{-\rho \cdot t} \cdot \ln C(t) dt = \int_0^{+\infty} e^{-\rho \cdot t} \cdot (\ln Y(t) + \ln(1 - s(t))) dt$$

The optimality is understood with respect to the utility function U represented by an integral with a discount coefficient ρ.

Application of the Pontryagin Maximum Principle

Hamiltonian function (Hamiltonian problem which measures the current flow of **utility** from all sources)

$$H(\theta, Y, \psi_1, \psi_2, s) = \ln Y + \ln(1 - s) + \psi_1 \cdot (s - a \cdot \theta) + \psi_2 \cdot a \cdot Y \cdot \frac{\theta}{\theta + 1}$$

Investment intensity that maximizes Hamiltonian function

$$\frac{\partial H}{\partial s} = -\frac{1}{1-s} + \psi_1 = 0 \Rightarrow s = 1 - \frac{1}{\psi_1} = \frac{\psi_1 - 1}{\psi_1}$$ (Investment intensity that maximizes utility)

2) $Y(t) = C(t) + I(t) = (1 - s(t))Y(t) + s(t)Y(t)$ ⇨ $C(t) = (1 - s(t))Y(t)$ ⇨ $\ln C(t) = \ln Y(t) + \ln(1 - s(t))$

where $C(t)$: consumption, $I(t)$: investment, and $s(t)$: investment intensity ($I(t)/Y(t)$).

(4) Hamiltonian System

Hamiltonian system with maximized s

$$H(\theta, Y, \psi_1, \psi_2) = \ln Y - \ln \psi_1 + \psi_1 \cdot \left(1 - \frac{1}{\psi_1} - a \cdot \theta\right) + \psi_2 \cdot a \cdot Y \cdot \frac{\theta}{\theta + 1}$$

(i) Price function (adjoint variable)

$$\dot{\psi}_1 = \rho \cdot \psi_1(t) - \frac{\partial H[\theta(t), Y(t), s(t), \psi_1(t), \psi_2(t)]}{\partial \theta} = \rho \cdot \psi_1(t) + a \cdot \psi_1(t) - a \cdot \frac{\psi_2(t) \cdot Y(t)}{(\theta(t)+1)^2}$$

$$\dot{\psi}_2 = \rho \cdot \psi_2(t) - \frac{\partial H[\theta(t), Y(t), s(t), \psi_1(t), \psi_2(t)]}{\partial Y} = \rho \cdot \psi_2(t) - a \cdot \frac{\psi_2(t) \cdot \theta(t)}{(\theta(t)+1)} - \frac{1}{Y(t)}$$

(ii) Cost function

$$\dot{Cost}(t) = \dot{C}_1(t) + \dot{C}_2(t) = \dot{\psi}_1(t) \cdot \theta(t) + \psi_1(t) \cdot \dot{\theta}(t) + \dot{\psi}_2(t) \cdot Y(t) + \psi_2(t) \cdot \dot{Y}(t)$$

$$\dot{C}_1(t) = \dot{\psi}_1(t) \cdot \theta(t) + \psi_1(t) \cdot \dot{\theta}(t)$$

$$\dot{C}_2(t) = \dot{\psi}_2(t) \cdot Y(t) + \psi_2(t) \cdot \dot{Y}(t)$$

(iii) Optimal control [→ (2)] [3)]

$$\dot{\theta}(t) = s(t) - a \cdot \theta(t) = 1 - \frac{1}{\psi_1} - a \cdot \theta(t) = 1 - \frac{\theta}{C_1} - a \cdot \theta$$

$$\dot{Y}(t) = a \cdot Y(t) \cdot \frac{\theta(t)}{\theta(t)+1}$$

(iv) Cost minimum

$$\dot{C}_1(t) = \rho \cdot C_1(t) - \frac{a \cdot \theta(t) \cdot C_2(t)}{(\theta(t)+1)^2} + \frac{C_1(t)}{\theta(t)} - 1$$

$$\dot{C}_2(t) = \rho \cdot C_2(t) - 1$$

Cost minimum condition
$$\Rightarrow \dot{C}_1(t) = \dot{C}_2(t) = 0 \Rightarrow C_2 = \frac{1}{\rho}$$

Solution of Stationary Equation of the Hamiltonian System $C_1 = \frac{\theta}{(1 - a \cdot \theta)}$ ➡ $\frac{(a+\rho) \cdot \rho}{a} \cdot C_1 = \frac{\theta}{(\theta+1)^2}$ ➡ $\frac{(a+\rho) \cdot \rho}{a} = \frac{1 - a \cdot \theta}{(\theta+1)^2}$

Solution of Stationary Equation for FD $\frac{(a+\rho) \cdot \rho}{a} = \frac{1 - a \cdot \theta}{(\theta+1)^2}$ ➡ $\frac{(a+\rho) \cdot \rho}{a} = \frac{(1 - a \cdot (FD - 1))}{FD^2}$ ➡ $\frac{(a+\rho)\rho}{a} \cdot FD^2 + a \cdot FD - (1 + a) = 0$

3) **FD maximum** ➡ $\dot{FD} = \dot{\theta} = 0$ ➡ $\theta = \frac{C_1}{a \cdot C_1 + 1}$ ➡ $\rho \cdot C_1 - \frac{1}{\rho} \cdot \frac{a \cdot \theta}{(\theta+1)^2} + \frac{C_1}{\theta} - 1$ ➡ $(a+\rho) \cdot \rho \cdot C_1 = \frac{a \cdot C_1}{a \cdot C_1 + 1} \cdot \frac{(a \cdot C_1 + 1)^2}{((a+1) \cdot C_1 + 1)^2}$

Normal form adjoint equation ➡ $\dot{\psi}_1(t) = \rho \psi_1(t) - \frac{\partial H}{\partial \theta}$, $\dot{\psi}_1(t) = \rho \psi_1(t) - \frac{\partial H}{\partial \theta}$ $(\psi(t) = e^{\rho t} \psi^*(t) \Rightarrow \dot{\psi}(t) = \rho \psi - \frac{\partial H}{\partial X};\ X = Y, \theta)$

ψ^*: steady state price of X satisfying the above condition.

References

Bauer, M. (1995). *Resistance to new technology*. Cambridge: Cambridge University Press.

Bell, D. (1973). *The coming of post-industrial society*. New York: Basic Books.

Hofstedo, G. (1991). *Cultures and organizations*. London: McGraw-Hill.

IEA (2008a). *World energy outlook 2008*. Paris: IEA.

IEA (2008b). International oil prices by WTI (West Texas Intermediate).

McDonagh, D. (2008). *Satisfying needs beyond the functional*. Paper presented at international symposium on the silver market phenomenon—business opportunities and responsibilities in the aging society.
METI (Ministry of Economy and Industry) (2008). Internal report on PV development.
METI (Ministry of Economy and Industry) (2009). Internal report on PV development.
Modigliani, F. (1968). *Life cycle hypothesis of savings, the demand for wealth and supply of capital*. Paper presented at Rome congress of Econometric Society.
Shin, J. H. (2009). *Sustainable functionality development in open Innovation toward a post-oil society*. Ph.D. dissertation, Tokyo Institute of Technology.
Watanabe, C. (1995). Identification of the role of renewable energy. *Renewable Energy*, *6*(3), 237–274.
Watanabe, C. (1996). Choosing energy technologies: the Japanese approach. In IEA (Ed.), *Comparing energy technologies* (pp. 105–138). Paris: OECD/IEA.
Watanabe, C. (1999). Systems option for sustainable development. *Research Policy*, *28*(7), 719–749.
Watanabe, C., & Asgari, B. (2004). Impacts of functionality development on the dynamism between learning and diffusion of technology. *Technovation*, *24*(8), 651–664.
Watanabe, C., & Zhao, W. (2006). Co-evolutionary dynamism of innovation and institution. In N. Yoda, R. Pariser & M. C. Chon (Eds.), *Chemical business and economics* (pp. 106–121). Tokyo: Chemical Society of Japan.
Watanabe, C., Griffy-Brown, C., Zhu, B., & Nagamatsu, A. (2002). Inter-firm technology spillover and the creation of a virtuous cycle between R&D, market growth and price reduction: the case of PV development in Japan. In A. Gruebler, N. Nakicenovic & W. D. Nordhaus (Eds.), *Technological change and the environment* (pp. 127–159). Washington: Resources for the Future (RFF) Press.
Watanabe, C., Moriyama, K., & Shin, J. (2009). Functionality development dynamism in a diffusion trajectory: a case of Japan's mobile phone development. *Technological Forecasting and Social Change*, *76*(6), 737–753.
Watanabe, C., Nasuno, M., & Shin, J. H. (2011). Utmost gratification of consumption by means of supra-functionality leads a way to overcoming global economic stagnation. *Journal of Services Research*, *11*(2), 31–58.
Watanabe, C., Shin, J. H., Heikkinen, J., Zhao, W., & Griffy-Brown, C. (2011). New functionality development through follower substitution for a leader in open innovation. *Technological Forecasting and Social Change*, *78*(1), 116–131.
Watanabe, C., Wakabayashi, K., & Miyazawa, T. (2000). Industrial dynamism and the creation of a virtuous cycle between R&D, market growth and price reduction. *Technovation*, *20*(6), 299–312.

Transition Towards Renewable Energy Supply—A System Dynamics Approach

Bo Hu, Armin Leopold, and Stefan Pickl

1 Introduction

There is no doubt that the resources which are daily consumed by modern industrial nations, like fossil fuels of different kinds or the storage capacity of atmosphere for CO_2 and other greenhouse gases, are finite. Also indisputable is that all national economies, the developed ones in particular, are all facing the need for substantial energy transitions. However, there seems to be no consensus about how to shape the structural conditions for this transition.

The electricity portfolio of the future will not focus any more on the two main categories of electricity, the base load and the peak load. The challenge that lies in store is to combine dispatchable, conventional (oil, coal and gas) and renewable but non-dispatchable (wind and photovoltaic) electricity power stations to a sustainable and reliable electricity supply portfolio.

In Germany the surely necessary target to reduce Greenhouse Gas emissions by about 40% in 2020 compared to 1990 is considered to be challenging, especially in combination with the decision to phase out nuclear power supply in 2022. The decision is strongly supported by the public. It should be assumed, however, that no substantial income loss or even economic down turn will be accepted in the context of the energy transition.

According to statistical data by the German Federal Ministry of Economics and Technology (BMWT 2011) wind and photovoltaic in Germany provide more than 23% of the capacity but less than 8% of the production in 2009 (Fig. 1). Moreover,

B. Hu (✉) · A. Leopold · S. Pickl
Universität der Bundeswehr Munich, Neubiberg 85577, Germany
e-mail: bo.hu@unibw.de

A. Leopold
e-mail: armin.leopold@unibw.de

S. Pickl
e-mail: stefan.pickl@unibw.de

J. Crespo Cuaresma et al. (eds.), *Green Growth and Sustainable Development*,
Dynamic Modeling and Econometrics in Economics and Finance 14,
DOI 10.1007/978-3-642-34354-4_10,

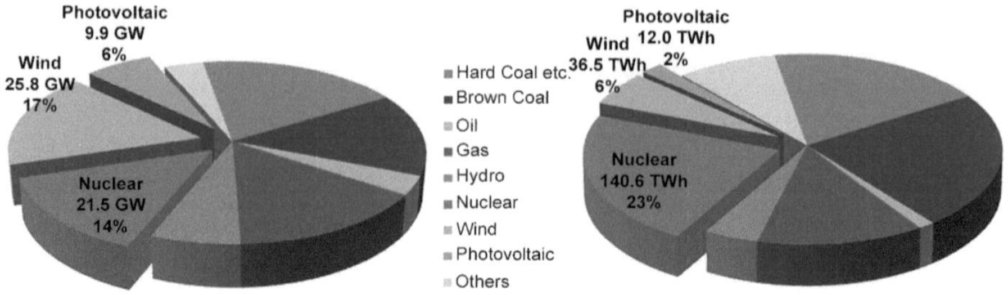

Fig. 1 Wind and photovoltaic in Germany provide more than 23% of the capacity but less than 8% of the electricity production in 2009 (see BMWT 2011) (color figure online)

despite the high potential of 380 GW of wind and photovoltaic production capacity of electricity in Germany (p. 63 in Umweltbundesamt 2010) both are not capable to deliver dispatchable or even continuous power supply. Nevertheless, due to the nature of wind and solar power there often exists a kind of temporarily electricity surplus. Therefore, along with 150 GW wind and photovoltaic and 65 GW dispatchable (including bio mass and hydro power) capacity in 2025, pumped storage power plants with a capacity of 30 GW and a electricity production of 10 TWh in Norway are seen to be a central part of the energy plan for Germany presented by the German Advisory Council on Environment (pp. 47–54 in SRU 2010).

Pumped storage is by far not the only way to store electrical energy and to turn the intermittent renewable sources into dispatchable ones (see, e.g., Chen et al. 2009). Especially with respect to the possible environmental impacts and costs of the necessary transnational high-voltage transmission cables and huge water storage reservoirs in Norway the conversion of renewable electricity into chemical fuels is a considerable alternative. Methane or hydrogen can be produced with the help of the Sabatier or electrolysis processes (see, e.g., Kolic and Clifford 1969; Lunde and Kester 1972). Hydrogen can be then converted together with CO_2 to synthetic natural gas (SNG). However, it has to be mentioned that because of the energy conversion processes only 30% to 40% of the originally generated wind power will reach the end consumer (see p. 18 in Sterner et al. 2011). Nevertheless, the main advantage of SNG is the fact that it can be directly stored and transported using existing gas supply systems.

In July 2010 Germany's Umweltbundesamt published a study which strengthens the role of chemical storage systems based on "eE-methane" or "eE-hydrogen" produced by renewable electricity (p. 37 in Umweltbundesamt 2010). In Germany the wind power stations are mainly located in the north while the large electricity consumers are located mainly in central and southern Germany. To reduce the load of the high-voltage transmissions cables in Germany renewable electricity can be used to produce methane or SNG which can be then transported via the already existing widely distributed gas pipeline system. Nowadays it is possible to enrich the natural gas with up to 10% methane according the latest regulation changes in the German electricity and natural gas supply law (Deutscher Bundestag 2005).

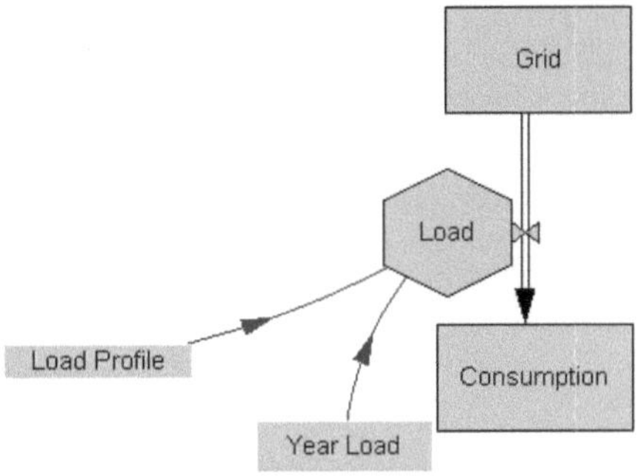

Fig. 2 Electricity is delivered from the grid (color figure online)

Based on our previous research in the area of emissions trading under uncertainties we are developing System Dynamics models which should depict the development of the energy market in a highly aggregated form and can be used to compare different possible pathways of the impeding energy transition. We use System Dynamics with the intention to make the modeling process more understandable while approaching the politically active public and to provide a transparent decision support method in regard to different energy concepts.

The preliminary results of the simulation runs using this model show that the SRU concept will only achieve 33% GHG mitigation in 2025 compared to 1990, despite the high costs due to planned huge storage capacity. A more effective GHG mitigation of about 40% can be achieved at lower cost by making use of higher wind and photovoltaic capacities in combination with the capability to produce synthetic natural gas using excess electricity from wind and solar energy.

In the following, we first introduce our System Dynamics model step by step using a serial of Stock-and-Flow diagrams in Sect. 2. In Sect. 3 we present and discuss some preliminary results of the simulations.

2 A System Dynamics Model for Electricity Supply

A System Dynamics model consists mainly of a number of interconnected stock and flow variables. A "stock is an accumulation, or integration, or level, to choose terminology from different fields. [A] flow changes the amount in [a] stock" (Forrester 2009). As shown in Fig. 2, electricity is delivered for `Consumption` from the `Grid`. Both are implemented as stock variables in our model and denoted as rectangles. The time specific consumption, `Load`, which is a flow variable and denoted as a hexagon, can be characterized not only by the `Year Load` but also by the `Load Profile` (in an hourly resolution). Notice that electricity cannot be stored without special facilities. The stock `Grid` has thus to be reset to zero at the beginning of each time step of the simulation.

Figure 3 depicts the `Dispatchable` power generation mainly using `Chemical Fuel` (or hydro power and geothermal power sources) on the one hand and non-dispatchable power generation using wind and photovoltaic (`WDPV`) energy on the other hand. The installed capacities are given by `Dispatchable Capacity`

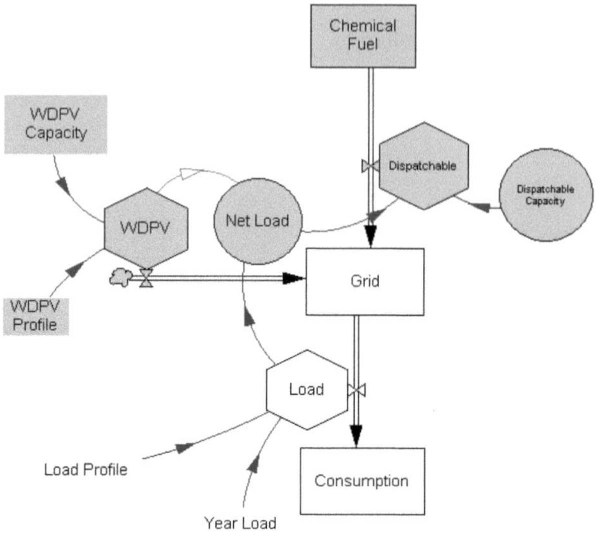

Fig. 3 Wind and photovoltaic as well as power plants using chemical fuels feed the grid (color figure online)

and `WDPV Capacity` respectively. `WDPV Profile` specifies the real wind and solar power generation in an hourly resolution. The `Net Load` or residual load is given by the difference of `Load` and `WDPV`. According to the notation we used in this work a blue and opaque arrow (f. i. from `Net Load` to `Dispatchable`) means a positive or concordant influence, whilst a red and transparent arrow (f. i. from `WDPV` to `Net Load`) depicts a negative or an opposite effect.

Notice that `Net Load` may sometimes be negative. The higher the `WDPV Capacity`, the more often `Net Load` is negative, and the more it makes sense to have the possibility to store electricity f. i. using pumped storage, as shown in Fig. 4. A storage system and its state are characterized by `Maximal Storage`, `Maximal Storage Power` and `Storage Efficiency` as well as `Stored Electricity`. As long as `Net Load` < 0 and `Filling Level` < 1 the storage is activated or `Do Store` > 0. Additionally, the storage is also activated if `Filling Level` < `Threshold`.

The stored electricity can be called up to provide grid stability when `Net Load` exceeds nearly the maximal `Dispatchable Capacity`. The `Call Power` is limited by `Maximal Call Power`. Additionally, the call function is characterized by `Call Efficiency` and a `Loss` of `Stored Electricity` caused by a technology specific `Loss Rate` has to be taken into account (Fig. 5).

As an alternative to storage, electricity can also be used to produce synthetic natural gas (`SNG`), as shown in Fig. 6. Depending on `SNG Capacity` and `SNG Efficiency` the net consumption of `Chemical Fuel` can be reduced.

Several additional parameters are used to complete the model (Fig. 7). First of all, the initial filling level of the storage is given by `Ini S`. The stability of electricity supply can be tested using `Stress Testing` which reduces `WDPV` and increases `Load` at the same time. The specific costs for fossil fuel and CO_2 emission permits are given by `Fuel M` (€/MWh) and `CO2 M` (€/tonCO2). The specific investment

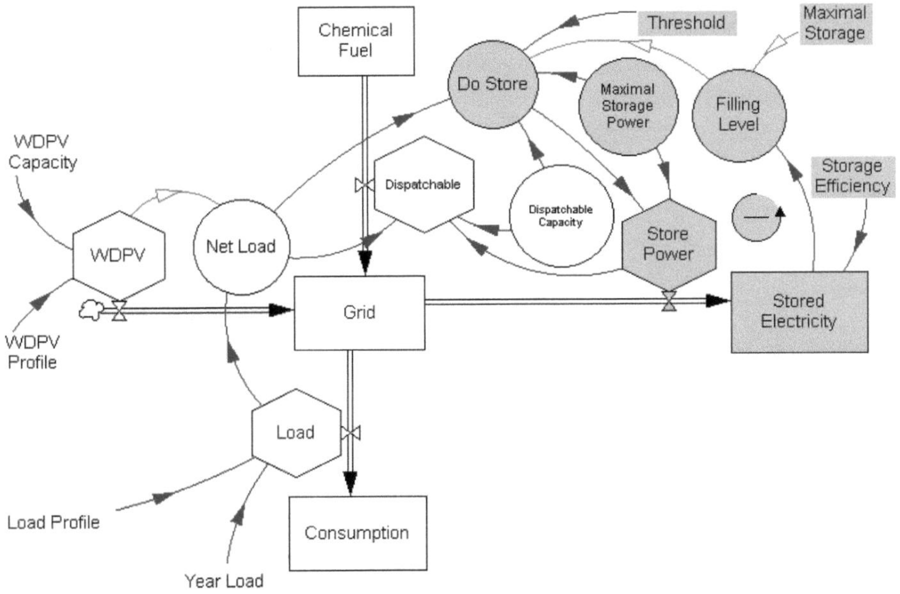

Fig. 4 Electricity can be stored f. i. using pumped storages (color figure online)

costs for storage are given by `Storage M` (B€/GWh), whilst the ones for WDPV, SNG and dispatchable power are given by `WDPV M`, `Dispatchable M` and `SNG M` (B€/GW).

3 Results of Simulations and Discussion

The Stock and Flow model discussed in Sect. 2 represents an integral equation system which can be solved using computational methods. In this way different electricity supply concepts can be presented by this model using different parameters and compared with each other regarding their reliability and resource consumption. To do this we first enter the characterizing key parameters of each concept, like `WDPV Capacity`, `SNG Capacity` and so on, and try to find the minimal `Dispatchable Capacity` which still provides reliable electricity supply under a given `Load Profile` and `WDPV Profile` for an entire year. A concept is considered as reliable if the cumulative energy shortage is less than 2.6 TWh during the entire year or 0.3 GW in average. Shortages are displayed in red color in the graph on the right side of our interactive user interface (Fig. 8). Notice that possible excess electricity occurring at another point of time does not offset the cumulative shortage in the calculation. In this way different concepts to be compared with each other are dimensioned on the same reliability level. The total production cost which includes investment, operating, fuel costs and emission permits is then calculated for each of the concepts.

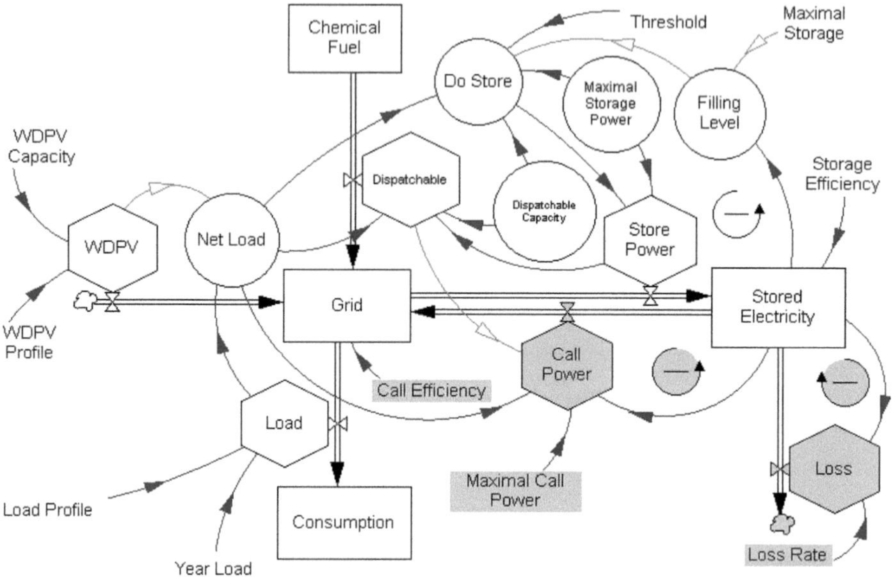

Fig. 5 High call power is necessary to compensate Net Load using stored electricity (color figure online)

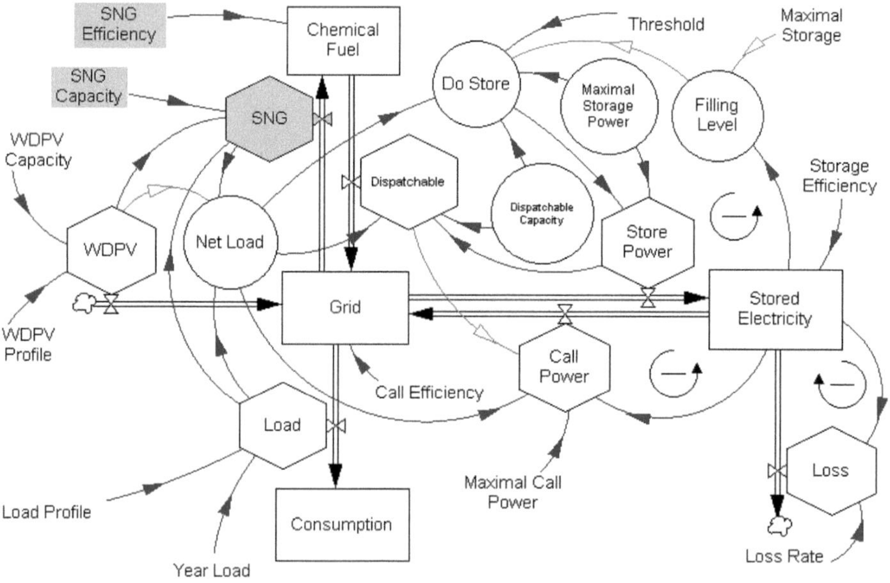

Fig. 6 One alternative option to pumped storage is to produce synthetic natural chemical fuels (color figure online)

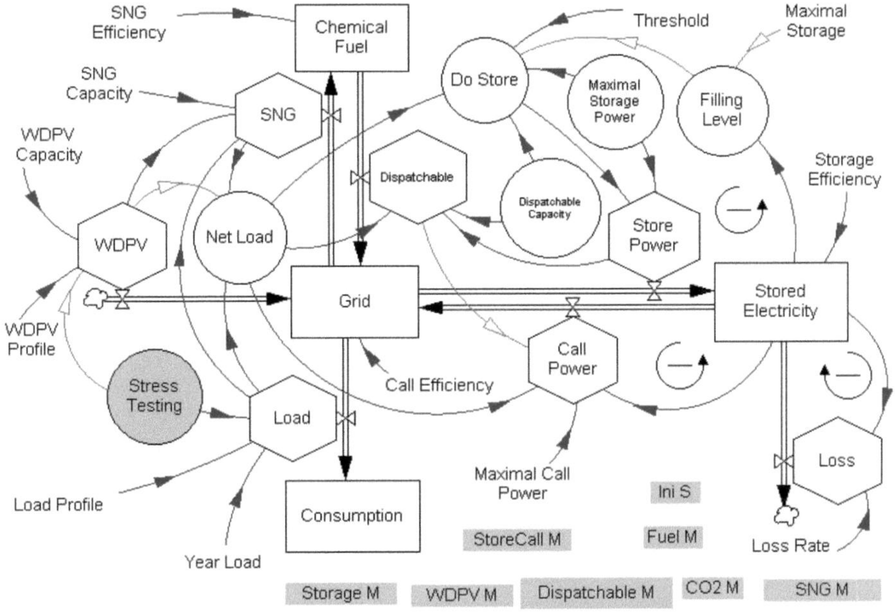

Fig. 7 Additional scenario parameters are included (color figure online)

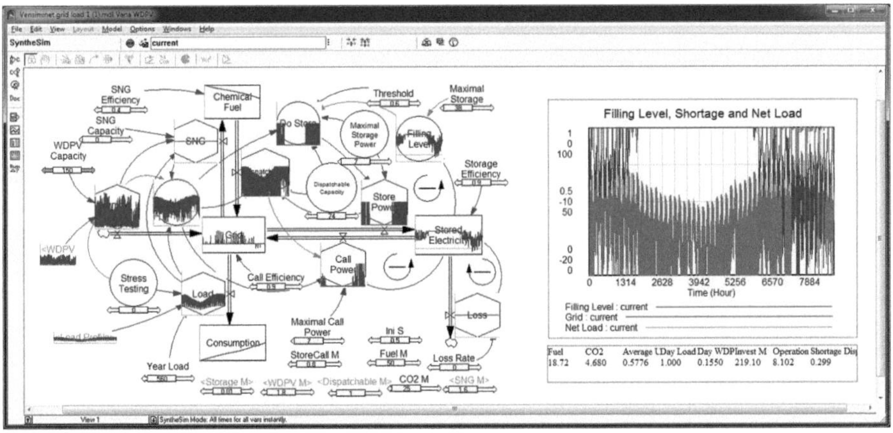

Fig. 8 Interactive user interface for simulations using Vensim PLE (Ventana Systems 2009) (color figure online)

Table 1 shows the parameters used for the simulations and the scenario-independent results of six different concepts for electricity supply. All six concepts are considered under three different price scenarios: (I) The prices of fuel and CO_2 permits remain the same, the annual interest rate for investment amounts to

Table 1 Parameters used for the simulations and scenario-independent results. Data sources for the estimated specific costs in the first column: BMWT (2011), Groscurth and Bode (2009), Reina (2008), SRU (2010), Sterner (2009), Umweltbundesamt (2010)

Spec. Costs	*Concept (2025)*		Cnv	Rnw	SRU	SNG	Rnw+	SNG+
	Year Load	TWh/a	560	560	560	560	560	560
1.0 B€/GW	Dispatchable Cap.	GW	82	74	60	74	72	72
1.8 B€/GW	WDPV Capacity	GW	36	150	150	150	200	200
0.8 B€/GW	Storage Power	GW	7	7	37	7	7	7
0.8 B€/GW	Call Power	GW	7	7	37	7	7	7
0.1 B€/GWh	Max. Storage	GWh	38	38	5500	38	38	38
1.6 B€/GW	SNG Capacity	GW	0	0	0	10	0	20
	CO2 Mitigation	%	6.5%	31.2%	31.3%	31.7%	38.4%	39.8%
	Shortage	GW	0.296	0.299	0.298	0.298	0.296	0.298
	Operation	B Euro/a	5.8	8.1	9.9	8.4	9.3	9.9
	Invest M	B Euro	21.9	219.1	307.8	235.1	307.1	339.1

Table 2 Comparative simulations between six different concepts and three price scenarios for 2025

Spec. Costs	*Concept (2025)*		Cnv	Rnw	SRU	SNG	Rnw+	SNG+
50 €/MWh_e	Fuel Cost	B Euro/a	25.47	18.72	18.71	18.60	16.78	16.39
25 €/tonCO2	CO2 Cost	B Euro/a	6.37	4.68	4.68	4.65	4.19	4.10
7.26%	Capital Cost	B Euro/a	11.5	25.8	32.2	27.0	32.2	34.5
Scenario I	Sum	B Euro/a	49.2	57.3	65.5	58.6	62.4	64.9
		€/MWh	88	102	117	105	112	116
100 €/MWh_e	Fuel Cost	B Euro/a	50.94	37.44	37.42	37.20	33.56	32.78
100 €/tonCO2	CO2 Cost	B Euro/a	25.47	18.72	18.71	18.60	16.78	16.39
7.26%	Capital Cost	B Euro/a	11.5	25.8	32.2	27.0	32.2	34.5
Scenario II	Sum	B Euro/a	93.7	90.1	98.2	91.2	91.8	93.6
		€/MWh	167	161	175	163	164	167
100 €/MWh_e	Fuel Cost	B Euro/a	50.94	37.44	37.42	37.20	33.56	32.78
100 €/tonCO2	CO2 Cost	B Euro/a	25.47	18.72	18.71	18.60	16.78	16.39
10.61%	Capital Cost	B Euro/a	16.8	37.7	47.1	39.4	47.1	50.5
Scenario III	Sum	B Euro/a	99.0	102.0	113.1	103.6	106.7	109.5
		€/MWh	177	182	202	185	190	196

6%; (II) Doubling of the fuel price and quadrupling of the price of CO_2 permits; (III) A higher annual interest rate of 10% on the basis of scenario II.

As shown in Table 2, it is obvious that the portfolio concept "Cnv" ("conventional") which does not include further expansion of renewable electricity supply

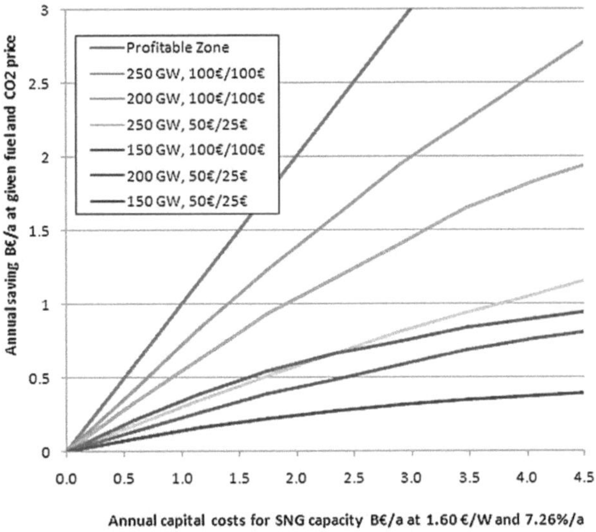

Fig. 9 SNG by its own is still far from achieving profitability (color figure online)

cannot provide the necessary CO_2 mitigation. It is thus not acceptable though it is the cheapest one in two of three scenarios. By contrast, the concept by SRU is the most costly one in all three scenarios because of its huge planned storage capacity. In spite of this it brings hardly any advantages in CO_2 mitigation compared to our reference concept "Rnw" ("renewable"). Both concepts fail to achieve the goal of 40% reduction. Even the concept "SNG" including SNG capacity misses the goal.

According to our simulations a combination of higher installed capacity of wind and solar power ("Rnw+") and synthesized natural gas production ("SNG+") seems to be the only concept which may achieve the goal of 40% CO_2 mitigation in 2025.

However, it has to be pointed out that from today's point of view the SNG as a single component is still far away from the profitable zone. It is hardly surprising that the profitability of SNG does not only depend on the specific investment costs and the conversion efficiency but also strongly on the utilization (therefore on installed wind and solar power capacity) and on the prices of fuel and emission permits, as shown in Fig. 9, since SNG technology is intended to convert exceed wind or solar powered electricity into chemical fuel to save fossil fuel and CO_2 emissions.

4 Conclusions

In this paper we describe the dynamics of the electricity supply in Germany using a System Dynamics model which focuses on the capacity of the four following subsystems: dispatchable conventional and non-dispatchable renewable (wind and photovoltaic) electricity supply, pumped storage in Norway as well as the production of synthetic natural gas (SNG) using excess (renewable) electricity. This model provides a transparent decision support method regarding the total cost and the GHG mitigation of different electricity supply concepts which are all dimensioned on the same reliability level.

Our simulations using different prices for energy and emission permits for the year 2025 show that a concept presented by the German Advisory Council on Environment (SRU) will only achieve about 31% GHG mitigation in 2025 compared to 1990, despite the huge costs due to the necessary storage capacity of 5500 GWh in Norway and the transport capacity of 30 GW between Norway and Germany. A more effective GHG mitigation of about 40% can be achieved at lower cost thanks to higher wind and photovoltaic capacities of 200 GW in combination with the capacity of 20 GW of the production of synthetic natural gas using excess electricity from wind and solar energy.

References

BMWT (2011). Stromerzeugungskapazitäten, Bruttostromerzeugung und Bruttostromverbrauch Deutschland. Bundesministerium für Wirtschaft und Technologie.

Chen, H., Ngoc Cong, T., Yang, W., Tan, C., Li, Y., & Ding, Y. (2009). Progress in electrical energy storage system: a critical review. *Progress in Natural Science*, *19*, 291–312.

Deutscher Bundestag (2005). Gesetz über die Elektrizitäts- und Gasversorgung.

Forrester, J. W. (2009). Some basic concepts in system dynamics. http://www.clexchange.org/ftp/documents/system-dynamics/SD2009-02SomeBasicConcepts.pdf (08.03.2010).

Groscurth, H.-G., & Bode, S. (2009). Anreize für Investitionen in konventionelle Kraftwerke – Reformbedarf im liberalisierten Strommarkt. arrhenius Institut für Energie- und Klimapolitik, Februar. 2009. http://www.arrhenius.de/uploads/media/arrhenius_DP2_Investionen_in_konventionelle_Kraftwerke.pdf (12.11.2011).

Kolic, E. S., & Clifford, J. E. (1969). Water-electrolysis cells using hydrogen-diffusion cathodes. Battele Memorial Institute. February 1969. http://www.dtic.mil/cgi-bin/GetTRDoc?AD=AD0691464 (05.05.2012).

Lunde, P. J., & Kester, F. L. (1972). Kinetics of carbon-dioxide methanation on a ruthenium catalyst. *Abstracts of Papers of the American Chemical Society*, *164*, 11–27.

Reina, P. (2008). Subsea cables tie European electric grids more closely. ENR.com, 01/23/2008. http://enr.construction.com/news/powerIndus/archives/080123.asp (06.05.2012).

SRU (2010). 100% erneuerbare Stromversorgung bis 2050: klimaverträglich. sicher, bezahlbar. Sachverständigenrat für Umweltfragen, SRU. http://www.umweltrat.de/cae/servlet/contentblob/1001596/publicationFile/63817/2010_05_Stellung_15_erneuerbare Stromversorgung.pdf.

Sterner, M. (2009). *Bioenergy and renewable power methane in integrated 100% renewable energy systems*. Dissertation, Universität Kassel, 23 September 2009. http://www.uni-kassel.de/upress/online/frei/978-3-89958-798-2.volltext.frei.pdf (06.05.2012).

Sterner, M., Jentsch, M., & Holzhammer, U. (2011). Energiewirtschaftliche und ökologische Bewertung eines Windgas-Angebotes. Fraunhofer-Institut für Windenergie und Energiesystemtechnik (IWES), Feb. 2011. http://www.greenpeace-energy.de/fileadmin/docs/sonstiges/Greenpeace_Energy_Gutachten_Windgas_Fraunhofer_Sterner.pdf (05.05.2012).

Umweltbundesamt (2010). Umweltbundesamt Deutschland: Energieziel 2050: 100 % Strom aus erneuerbaren Quellen.

Ventana Systems (2009). Free download | Vensim PLE. http://www.vensim.com/freedownload.html (19.07.2009).

MIX
Papier aus verantwortungsvollen Quellen
Paper from responsible sources
FSC® C105338

If you have any concerns about our products,
you can contact us on
ProductSafety@springernature.com

In case Publisher is established outside the EU,
the EU authorized representative is:
Springer Nature Customer Service Center GmbH
Europaplatz 3, 69115 Heidelberg, Germany

Printed by Libri Plureos GmbH
in Hamburg, Germany